KU-575-019

MANUFACTURING RESEARCH AND TECHNOLOGY 1

FLEXIBLE MANUFACTURING

Recent Developments in FMS, Robotics, CAD/CAM, CIM

MANUFACTURING RESEARCH AND TECHNOLOGY 1

Flexible Manufacturing

Recent Developments in FMS, Robotics, CAD/CAM, CIM

Edited by

A. Raouf

Department of Systems Engineering, University of Petroleum and Minerals, Dhahran, Saudi Arabia

S.I. Ahmad

School of Computer Science, University of Windsor, Windsor, Ontario, Canada

ELSEVIER
Amsterdam — Oxford — New York — Tokyo 1985

ELSEVIER SCIENCE PUBLISHERS B.V.
1 Molenwerf,
P.O. Box 211, 1000 AE Amsterdam, The Netherlands

Distributors for the United States and Canada:

ELSEVIER SCIENCE PUBLISHING COMPANY INC.
52, Vanderbilt Avenue
New York, NY 10017

ISBN 0-444-42504-7 (Vol. 1)
ISBN 0-444-42505-5 (Series)

© Elsevier Science Publishers B.V., 1985

All rights reserved. No part of this publication may be reproduced, stored in a retrieval system or transmitted in any form or by any means, electronic, mechanical, photocopying, recording or otherwise, without the prior written permission of the publisher, Elsevier Science Publishers B.V./Science & Technology Division, P.O. Box 330, 1000 AH Amsterdam, The Netherlands.

Special regulations for readers in the USA — This publication has been registered with the Copyright Clearance Center Inc. (CCC), Salem, Massachusetts. Information can be obtained from the CCC about conditions under which photocopies of parts of this publication may be made in the USA. All other copyright questions, including photocopying outside of the USA, should be referred to the publisher.

Printed in The Netherlands

PREFACE

Before the Industrial Revolution, firms that manufactured goods were small, loosely organized and their skilled craftsmen made an entire product by hand. Improvements in transportation and communication in the 19th century resulted in mass markets that demanded production volumes for which the preindustrial firms were unsuited.

Frederick W. Taylor's technique of breaking work into individual elements that were simple, well defined, highly specialized functions, which could be performed repetitively by a machine or a worker, yielded the greatest production results. Most of the companies using this technique were engaged in making large quantities of the same thing. This type of production system is called a "Dedicated Manufacturing System". To be profitable, such systems must always run at full capacity and make a single product that requires no change.

In the late 1960's, consumer preference was for greater diversity and more customized features of products. Production systems that were able to satisfy the consumers' preferences offered intense competition to the dedicated production systems. Such systems came to be known as "Flexible Manufacturing Systems". In these systems, machines capable of performing a wide and redefinable variety of tasks were substituted for machines dedicated to the performance of specific tasks. The time required to effect design changes in a Dedicated Manufacturing System invariably is extremely high. Flexible Manufacturing Systems can also be programmed to handle new products thus extending the machines' life cycles. Flexible Manufacturing Systems represent a change from "Standardized goods produced by customized machines" to "customized goods produced by standardized machines".

Scholars and researchers are showing an ever-increasing interest in attending to problems related to Flexible Manufacturing Systems. Such interest led to the VIIth International Conference on Production Research which was held in Windsor, Ontario, Canada during 1983. With a view to provide a ready reference and a sort of state-of-the-art, the selected papers pertaining to Flexible Manufacturing Systems presented at this conference have been assembled in one volume. This book will be of much interest to researchers, managers and students of Flexible Manufacturing Systems and allied areas. Due to constraints of time, editing has been kept to a minimum. The errors that remain are considered to be less important than having the book available sooner.

We are grateful to the authors for allowing us to prepare this collection. Our thanks are also due to those who have assisted us in the work of publication.

Abdul Raouf

S.I. Ahmad

CONTENTS

PART TWO
ROBOTICS AND CAD/CAM

INTRODUCING A FLEXIBLE MANUFACTURING SYSTEM

A.S. CARRIE[1], E. ADHAMI[1], A. STEPHENS[1] and I.C. MURDOCH[2]

[1] Dept. of Production Management and Manufacturing Technology, University of Strathclyde, Glasgow, Scotland.

[2] Anderson Strathclyde, Glasgow, Scotland.

ABSTRACT

Flexible Manufacturing Systems represent a significant investment and their introduction involves a major project for a company. This paper refers to the introduction of FMS in a company. It discusses various management aspects of the project, and its main subject is the simulation modelling of the system. The objectives of the simulation studies are given and the model described. Results are given and commented on, whereby certain capacity limitations were identified. As a result of the studies, weaknesses in the supplier's control software were highlighted and modifications made.

INTRODUCTION

A Teaching Company Scheme between the University of Strathclyde and Anderson Strathclyde plc was set up in 1977. Among the objectives were the improvement of performance by accelerating the Company's implementation of new technologies in both manufacturing technology and information processing techniques. Main achievements include wider use of the cell layout system, more computerization of production control, and the introduction of turnkey systems for CAD/CAM. This paper deals with perhaps the most imaginative of these innovations, the introduction of Flexible Manufacturing System for the manufacture of large castings.

THE COMPANY AND THE PRODUCT

Anderson Strathclyde plc is a Scottish-based engineering company. It employs some 4500 employees and has a turnover of about £100 m. The company manufacture coal-mining machinery. The products are sold throughout the world and can be found in most coal-mining countries: international competitiveness is therefore important. Its main manufacturing locations are in the U.K. but it has subsidiaries in the U.S.A., Australia and South

Africa. The largest factory, where the FMS is to be installed, is at Motherwell and employs about 730 direct workers.

The largest machine in the range is the AM500 Double-Ended Ranging Drum Shearer which weighs 40 tons, measures 40 feet in length and can cut coal at a rate of 1.5 million tonnes per annum. The machine comprises several interchangeable modular units such as motor gearbox, power pack, and gearhead, which are selected to suit the needs of individual coal faces. The units are housed in large steel castings which represent the largest value category of manufactured items, with about 1000 produced annually, costing around £2500 in the raw state, and weigh up to 2.5 tons. Their manufacture is a complex process involving approximately 200 to 300 tools and took some months to machine. The overall lead time from ordering the castings to delivery of the machine is about 9 months, but since the market demands delivery in 4 months, the production programme must be initiated on a speculative basis 5 months before sales can be expected. The principal machine used in producing these castings is the horizontal boring machine. The company has 40 of these machines, but the workload was enough for about 65 double-shifted (45 on these major castings and 20 on other work) so there was a continual need for subcontract capacity.

In response to these problems of lead time and capacity the company had introduced two CNC horizontal boring machines about four years ago. Significant results had been achieved, with lead times reduced from 6 months to 3 and boring and milling work reduced from 160 hours to 80. The company could have invested in more of these machines, but since the number of tools required was two to three times the tool magazine capacity problems of tool changing, setting and handling would remain. Consequently the FMS approach justified exploring.

THE FLEXIBLE MANUFACTURING SYSTEM

Two companies undertook design studies, with the cost underwritten by the Teaching Company Scheme. One company declined to submit an FMS proposal due to the complexity of the workpieces, but the other, Giddings & Lewis Inc., did propose an FMS through their Scottish subsidiary, Giddings and Lewis-Fraser.

This proposal was for a system of eight machines which would produce castings in batches in sets of gearboxes and booms (to provide a mix of work). This was subsequently amended to a system of six machines, which can process castings in a completely random mix and sequence. An order was placed in September 1981, commissioning will begin in summer 1983 and the system will build up to full capacity in 1984. The layout is shown in Fig. 1.

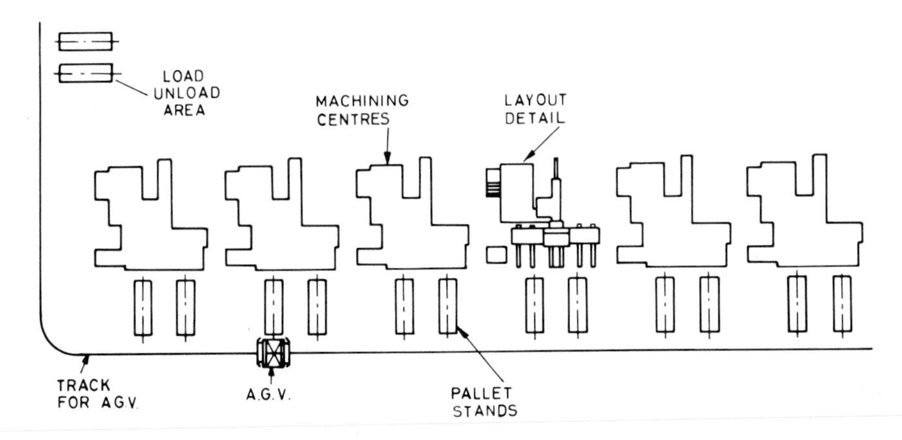

FIG. 1 LAYOUT OF FLEXIBLE MANUFACTURING SYSTEM

The system will comprise five live spindle CNC horizontal boring machining centres and one borer with facing head, all of which will have a tool changer with a hundred tool capacity magazine. Castings will be fixturized and moved on pallets by an Advanced Guided Vehicle (AGV) following a wire buried in the floor. At the load/unload area and at each machine there will be two pallet stands acting as buffers between the AGV and the machine table. There will be thirteen pallets. The whole system will be controlled by a DEC PDP 11/44 executive computer which will route parts through the system, schedule machining operations, monitor tool life and store part programs. It will also produce management reports on system status, machine status, component status, tool status and maintenance and failure diagnostics.

The objectives of the system are:

- to reduce throughput time in the factory
- to decrease speculative inventory
- to reduce product cost
- to maintain or improve quality
- to automate handling and eliminate manual operations or handling to the maximum degree
- to handle batches of one-off
- to accept design alterations and new products with minimum re-tooling

The financial justification of the FMS was based on discounted cash flows over a ten-year period. The gross costs of the system are: capital £6.2 m and revenue £1.3 m. Under the capital heading are the purchase of the machines, the AGV, the computer and site preparation, while under revenue are costs of relocating existing plant, subcontracting during relocation, salaries and expenses of project team, training and commissioning costs. However, after applying Regional Development Grant, Department of Industry Grant and European Social Fund the net expenditure becomes: capital £2.5 m, revenue £1 m, totalling £4.5 m.

The move from conventional to CNC machining had reduced both lead times and machining times by 50% and with the FMS it is expected that lead times will be reduced to one-sixth and machining times to one-fifth of the original figures. Based on producing 1000 castings per annum this will save the equivalent of 30 men and take £1 m out of inventory. The financial savings were sufficient to make the project viable, without considering hidden benefits such as consistency of manufacturing time, improved quality, reduction of human element, and reduction of number of castings in progress from about 400 to 130.

A network diagram for installing the system is shown in Fig. 2. Main activities or groups of activities are concerned with

relocation of machines in building
determining manning levels, selection and training of personnel
preparation of the building
manufacture, delivery, installation, testing of machines
specification, manufacture, installation of AGV
specification, writing and testing of software
installation of computer
determining, ordering and delivery of consummable tools, durable tools, fixtures, and racking needed for them
determining, ordering and installing swarf removal plant

Although not shown, durations were estimated for each activity and the network revised regularly to reflect progress achieved.

One of the significant decisions concerned the location of the system. Some argued that in view of the special nature of the system it should be located in a new building. However, it was decided to locate the system within the building where this work has been done up till now in order to emphasize that the system is just another production facility and to avoid creating 'mystique', and also to minimize capital costs.

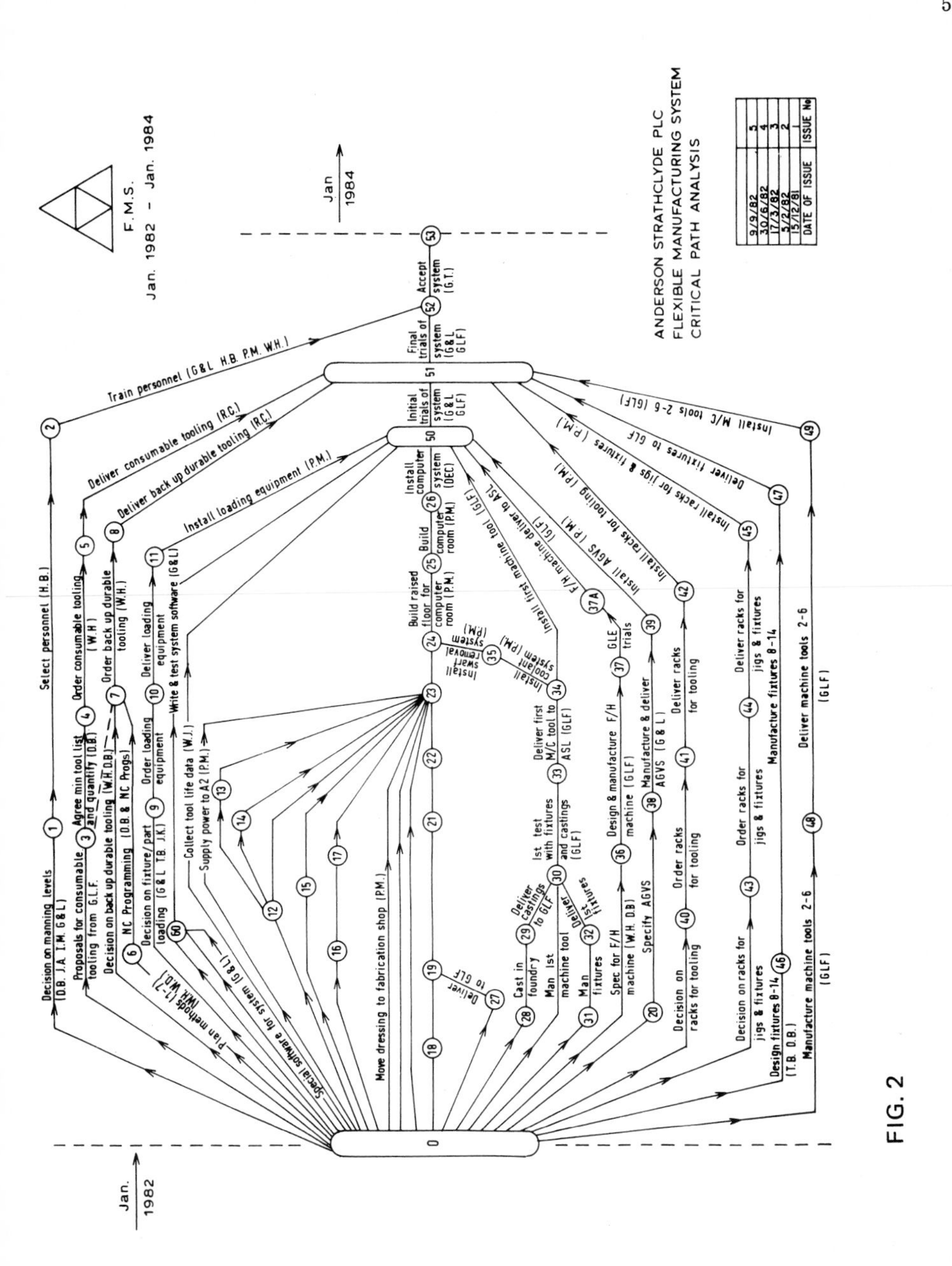

FIG. 2

SIMULATION STUDIES OF THE SYSTEM

With an investment of this magnitude, a company wants to be certain that target levels of output will be achieved. At the tendering stage firm operational data is in short supply, for such simple reasons as that operation time figures are only estimates until detailed part programming has been done. Consequently, management also requires to know how sensitive will be the operating performance of the system to variations in part mix, operation times, machining methods and so on. Simulation is an appropriate method of evaluating these parameters, and in this case the manufacturer presented simulation results to support his system design and performance predictions. It was therefore planned to undertake simulation studies for the following objectives:

to verify predictions given by the manufacturer

to assess the sensitivity of system performance to variations in part mix, operation times, facility reliability, etc. or the incorporation of extra facilities such as a robot vacuum cleaning station

to examine priority rules for launching parts into the system, operation sequencing, materials handling and so on, so that day to day operating policies for the real system could be developed

to assess the implications of putting other parts on the system as new parts are designed or current parts are replanned for FMS manufacture

Company management attached considerable importance to this work, and had written into the acceptance terms that the system could meet the simulated performance.

The model

The model was written using the Extended Control and Simulation Language (Clementson 1982). This is an activity-based language in which the Activity Cycle Diagram of the system to be modelled is defined. This diagram describes the lift story of each entity in the system. Figure 3 shows the part of the activity cycle diagram for the load/unload area, and Figure 4 the part for the machine area.

Castings arrive (activity Arrive) from outside (queue Outside) the system as raw castings and wait for service (queue Wait). First the casting is put into a fixture at a work bench with the aid of an overhead crane (activity Put on Fixture) and is then ready (queue Ready) to be placed (activity Place) on a pallet at one of the two pallet stands at the load/unload area. Once placed on a pallet the casting and pallet move together through the system and to simplify the model the casting is regarded as 'in the system' (queue Insystem) until brought back on its pallet to the load/unload area. Then it is removed from the pallet

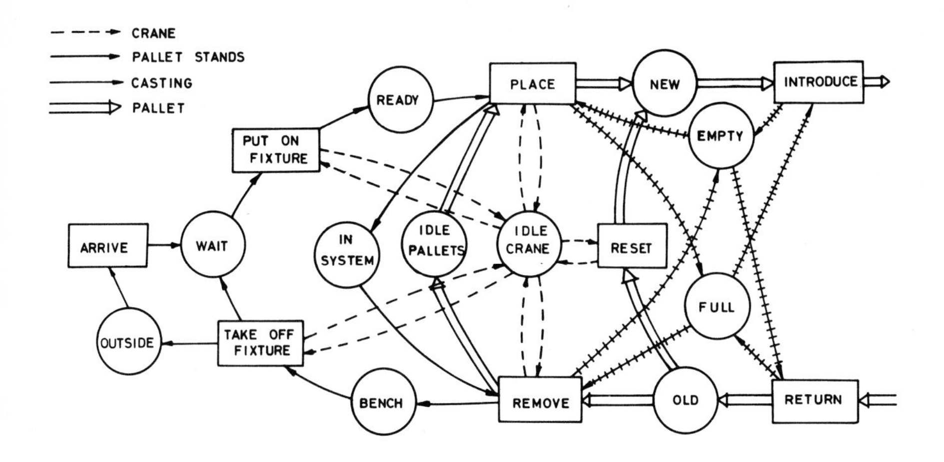

FIG. 3 ACTIVITY CYCLE DIAGRAM AT LOAD/UNLOAD AREA

(activity Remove), placed on a work-bench where it waits (queue Bench) to be taken out of (or off) the fixture (activity Take off Fixture). After being taken off its fixture the casting may be completely machined in which case it is taken away to the assembly area (queue Outside) or else it may require further machining on a new fixture, in which case it goes into queue Wait.

Fixture availability must be checked before castings can be put on fixtures. This is done by maintaining a variable defining how many fixtures of each type are available. Castings can only be

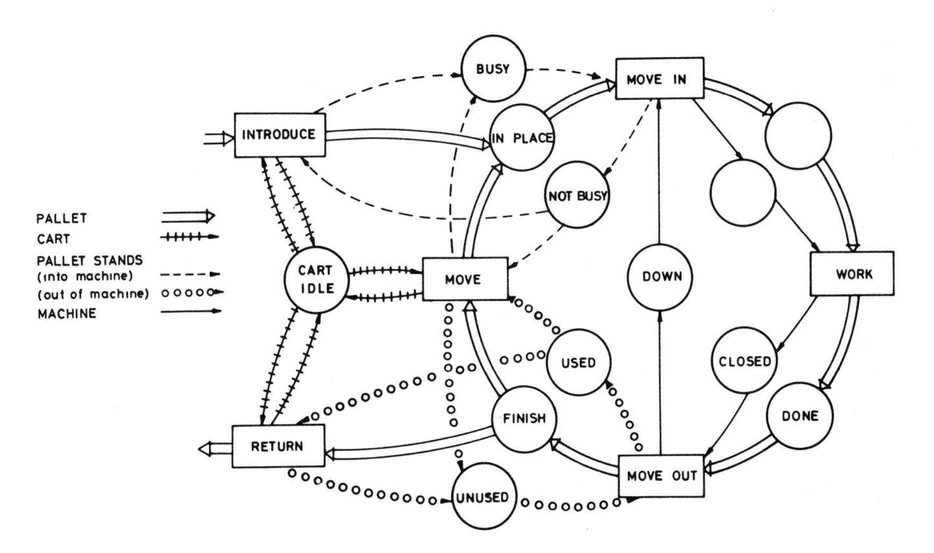

FIG. 4 ACTIVITY CYCLE DIAGRAM FOR MACHINE AREA

put on a fixture if this quantity for the required type of fixture is greater than zero. The number of available fixtures is decreased by one in activity Put on Fixture and increased by one in activity Take off Fixture. By this means the formal treatment representation of the activity cycle for fixtures can be excluded from the diagram and model.

A pallet, once a casting had been placed on it, waits (queue New) to be moved to a machine (activity Introduce) by the AGV. The pallet is placed on the pallet stand at a suitable machine (queue In Place). When the machine is ready to process the casting the pallet is moved on to the machine table (activity Move in), then machined (activity Work). The pallet may have to wait (queue Done) before being moved from the machine (activity Move out) on to the pallet stand, where it waits (queue Finish) to be moved away by the cart. The move may be direct to another machine (activity Move) or it may be back to the load/unload area (activity Return). When a pallet has been brought back to the load/unload area it waits on the pallet stand (queue Old) for service. Either the part and its fixture are lifted off the pallet (activity Remove) or the casting may be re-orientated within the current fixture (activity Reset).

All the moves by the cart require that the destination pallet stand is unoccupied. One of the two pallet stands at each machine is regarded as the queue position for pallets awaiting machining and the other is regarded as the queue position for pallets waiting to be moved elsewhere after machining. Thus activities Move and Introduce take a pallet stand from queue Not Busy and put it into queue Busy. Activity Movein returns the pallet stand from Busy to Not Busy. Similarly activity Moveout converts a wayout pallet stand from Unused to Used, while Move and Return convert it from Used to Unused.

At the load/unload area it is not necessary to treat the pallet stands as wayins or wayouts. Returning a pallet to the area converts a stand from Empty to Full, while activity Introduce does the opposite.

The activity cycle for the machining centres appears in Figure 4. From activity Movein to activity Moveout it is similar to the cycle for the pallet. Between Moveout and Movein the machine passes through the queue Down, awaiting the arrival of another pallet.

The operators at the load/unload area have not been included since all the relevant activities (Remove, Take off Fixture, Put on Fixture and Place) also involve the overhead crane and since there is only one of these, this was taken as the limiting factor.

The operation sequence of a typical casting involves three stages of machining using a different fixture at each stage. During the first stage most castings have to return the load/ unload area to be reset without changing the fixture. The machining operations at each stage may involve roughing, roughing and finishing, or finishing and facing head machines. The facing head machine is used for operations involving manual intervention. Most of the castings require back-facing operations, in which the cutting tool has to be inserted manually into the holder after the holder on the spindle has been positioned through a bore beyond the face to be machined.

This model implies that empty pallets are held off the system at the load/unload area. This is a simplification, because in reality pallets will remain in the system whether loaded or empty. However this model was important because it provided a yardstick for comparison with the manufacturer's results since his model made a similar assumption. A revised version explicitly included the movement of empty pallets around the system and various rules for determining when and where they should be moved.

Results

A most important result was that the system could become blocked. No pallets could be moved, because the in-queue position of the required machine was occupied. The casting there could not move on to the machine because the table was occupied. The casting on the table could not be moved out because the out-queue position was occupied. Nor could the casting at the out-queue position be moved, since it wanted to go to a machine where a similar situation occurred. Logic had to be added to the model to handle this problem.

Empty pallets. Run 1 is from the model in which empty pallets were removed from the system, while Run 2 is from the model which included empty pallets. This shows that the AGV is busy about 25% longer to move empty pallets, with a consequent reduction in system output of almost 10% and reductions in machine utilization.

Variations in operation times. When the production tapes were being prepared it was decided that an additional roughing

	Run no.									
	1	2	3	4	5	6	7	8	9	10
Production rate (parts/week)	18.5	17.1	14.5	10.5	16.8	16.5	20.1	18.3	17.9	17.2
Mean flow time (min.)	2631	2715	2852	4418	2780	2787	2216	2605	2694	2781
Utilization of:										
Roughing machines	84	80	96	52	78	76	91	84	82	80
Finishing machines	65	61	52	38	60	59	72	65	63	61
Facing head machine	83	78	65	97	76	75	90	83	81	78
AGV	35	44	37	29	42	43	44	35	34	33
Crane	24	22	19	14	22	23	25	24	23	23
Pallets	55	53	55	54	55	52	81	58	56	54
Fixtures	36	37	50	31	43	38	33	39	37	35

Results of simulation runs.

The table summarizes the results obtained from ten runs, examining senstitivity to various factors.

operation might be needed. Run 3 shows that the effect of a 75% increase in roughing time is a reduction in production rate of 15% due to very high utilization of roughing machines. The times for facing head operations were approximate estimates and company engineers feared the supplier may have seriously underestimated the work content of these operations. To assess this, in Run 4 these times were increased by 100%. This shows a substantial drop in output relative to run 2 of 60% and the facing head machine becoming a serious bottleneck. To relieve this in practice an extra shift might need to be worked on this machine alone.

Vacuum swarf removal. Swarf removal is a practical problem, since many chips can remain inside the casting. This can interfere with subsequent operations, and certainly must be removed before assembly. It was proposed to include in the system a robot vacuum cleaning station where a vacuum tube, controlled by a robot, could be inserted inside the casting to remove lodged chips. Three possible locations are: at the load/unload area, at a separate station while the casting is on the AGV or at a separate station on an extra pallet stand. Run 2 included vacuuming being done on the AGV. Run 5 assumes the existence of the additional pallet stand. Run 6 assumes vacuuming done at the load/unload area. The system is relatively insensitive to these changes.

Fixture availability. The fixtures are expensive and purchasing only one set of each is planned. However, it is important to know whether this will restrict the output of the system. Run 7 assumes the availability of two sets of each fixture and shows improvement in output of about 18%.

Reliability. The company already have several horizontal boring machines by this manufacturer and are satisfied with their reliability. The manufacturer's simulation assumed that the system would be up for 75% of its available hours and down for 25%. The company considered this a reasonable assumption so it was incorporated in the models. However, the reliability of pallet stands was an unknown factor. Runs 8, 9 and 10 incorporate failure of a pallet stand at a roughing machine, a finishing machine and the facing head machine respectively after 3 weeks for the remaining 10 weeks of the simulation run. Logic had to be incorporated avoiding the restriction that one stand was the way-in queue and one was the way-out. Because of this complication these runs exclude empty pallets. Comparison with run 1 shows slight reductions in output depending on the pallet stand concerned.

CONCLUDING REMARKS

Apart from the basic requirement to have the FMS installed, tested and accepted, there have been many questions concerning predicting the production rate which the system will achieve, and of determining how the best productivity can be achieved.

During the time between the proposal and acceptance many things can change. The parts which will be machined have been altered. The initial time estimates for the operations have been revised as the methods were studied more closely. These revisions have emerged over the period, so frequently simulation runs had to be repeated with new data. The simulation runs raised several questions which had to be referred back to the supplier for comment. The blockage problem in particular raised questions of how the system will cope.

The company have made various requests concerning the executive software, mainly about facilities for performing scheduling work or linking with other production control computers. These have been found possible or otherwise over a lengthy period. Not all that the company would like has the supplier been able to provide. The simulation models may be developed to provide a short-term scheduling tool. This will require data on tooling and current status to be available involving a very significant increase in the volume of data to be handled. This will become more important in the near future as more parts are planned for production on the system. The tool magazine capacities are almost exhausted by the seven parts being planned for the system initially. It may become necessary to control the mix of parts in the system at any time in order not to require too many tools.

There have been no industrial relations problems in the company, in contrast to that in the supplier where lay-offs and their effects caused some delay to the project.

It is too early to comment on the eventual pay-off of the system. However, one can say that if the predicted values are achieved the FMS will achieve an order of magnitude improvement in productivity.

REFERENCES

Clementson, A.J., 1982, E.C.S.L. User's Manual (Birmingham: Clecom Ltd).

Lenz, J., 1980, M.A.S.T. User's Manual (Oshkosh, Wisconsin: CMS Research Inc).

FLEXIBILITY IN PULL AND PUSH TYPE PRODUCTION ORDERING SYSTEMS - SOME WAYS TO INCREASE FLEXIBILITY IN MANUFACTURING SYSTEMS

R. MURAMATSU[1], K. ISHII[1] and K.TAKAHASHI[1]
[1]Department of Industrial Engineering, School of Science & Engineering, Waseda University, 3-4-1 Ohkubo, Shinjuku-ku, Tokyo 160 (Japan)

ABSTRACT

In recent years, many flexible manufacturing systems have been developed (ref.1,2,3,4). FMS is important in order to adapt to severe changes in market conditions and technology and to increase productivity.

This paper introduces variables for evaluating flexibility and discusses ways to improve the flexibility of single-stage production systems with different characteristics. However, since most production systems are multi-stage mixed type systems, this paper also shows how the "amplification" or increase in the variability of multi-stage production systems is an important factor affecting flexibility and productivity. Ways of reducing the "amplification" by using production ordering systems are also presented in this paper.

The multi-stage production ordering models discussed here are simplified. They neglect characteristics of single production systems which comprise total multi-stage production systems. The models also neglect transportation and material flow parameters.

The flexible manufacturing system of today is treated as one stage which is part of an overall system consisting of a succession of production stages with different characteristics. However, it should also be thought of as a way to increase the flexibility and productivity of the system as a whole.

INTRODUCTION

There are two major areas for consideration by management.
The first is the marketing area. Cultivation of new markets and changes in user needs require new products. Market segmentation increases the number of product specifications. Increases in the number of products create greater fluctuation in each product's demand, and product life cycles grow shorter.

The second is the production area in which product lines and specifications are varied, parts configurations are complex and the precision and speed of the production equipment in each stage differ according to changes in technology.

Moreover, most industries have many successive stages of production, inventory and material handling. In these cases, the higher the stage, the longer the total lead time required from the material processing stage to the final assembly line.

MANAGEMENT NEEDS TO ADAPT PRODUCTION SYSTEMS TO MARKET AND TECHNOLOGY CHANGES

In order to adapt production systems to changes in the market and production areas, production management must consider the following:

1) Minimization of total lead time from material processing to shipping and minimization of total cost.

2) Minimization of time needed to change to new products.

3) Keep down the amount of lead time and keep the inventory at the same level in each stage, regardless of increases in the variety of products.

4) Keep variations in total work load and inventory at the same level even if the fluctuation in demand for each product increases.

5) Keep variations in the production ordering and inventory levels the same in the face of forecasting error and down time.

6) Keep the "amplification" in the production ordering and inventory level in preceding processes the same.

7) Control production factors which increase total cost.

PRODUCTION SYSTEMS TO SATISFY MANAGEMENT'S NEEDS AND INCREASE FLEXIBILITY

There are many ways to satisfy management's needs for each production type. There are also many different opinions regarding the concept of "flexible manufacturing systems" (refs. 5-12).

The concept of the "flexible manufacturing system" is defined in this paper as a system which satisfies management's need to minimize resources and time as mentioned above. The variables used to evaluate flexibility are the number of parts and products, the length of total lead time, quantity of inventory and the "amplification" in production and inventory level in each stage of the system under the constraints of investment and product costs.

In this paper, the production types are classified into single production systems and multi-stage production systems. Further, single stage production systems are classified into machining process systems, lot production systems and assembly line production systems according to differences in production scheduling, equipment and production methods. The multi-stage production systems are classified into multi-stage mixed types of machining processes, multi-stage lot production systems and multi-stage machining, lot and assembly line production systems.

FACTORS INCREASING FLEXIBILITY IN SINGLE STAGE PRODUCTION SYSTEMS

In single stage production systems, there are many factors which increase flexibility within each production type.

In the machining systems, in order to adapt to a variety of parts or products, it is necessary that the set up time(t_s) and the machining time(t_m) are shortened for minimizing lead time. And it is necessary that the feeding and moving out time(t_h) are shortened for minimizing inventory quantity. Thus, realizing investment saving depends on the use of alternative equipment. Labor cost savings may be realized by automation and robotization.

Adapting lot production systems to handle an increased variety of parts or products requires shortening the set up time and making smaller production lot sizes(q). Set up time(t_s) and processing time(t_m) must also be shortened in order to make larger production lot sizes which minimize lead time. It is also necessary to shorten set up time to minimize inventory quantity. The realization of investment savings depends on the use of alternative equipment. Labor costs saving can be realized through automation and robotization.

Assembly line production systems change to multi-model mixed assembly line production systems in order to adapt to a variety of products and to minimize the total lead time and inventory level required. Investment in the systems can be reduced through equipment overlapping. Costs can be reduced through inventory savings. Labor costs would be saved by increasing the ratio of versatile operators (ref. 13).

Factors increasing the flexibility in the single stage production systems are shown in Table 1.

FACTORS INCREASING FLEXIBILITY IN MULTI-STAGE PRODUCTION SYSTEMS

Concerning the flexibility of multi-stage production systems, the key problem is to prevent an increase of "amplification" in production ordering and inventory level from final production stage to the stages preceeding final production (ref. 14).

The "amplifications" are defined by equation (1).

$$\begin{aligned} \mathrm{Amp}(O^k) &= V(O^k)/V(D) \\ \mathrm{Amp}(B^k) &= V(B^k)/V(D) \end{aligned} \quad (1)$$

where, $V(O^k)$:quantity variance upon production ordering at the k-th stage,

$V(D)$:market demand variance,

$V(B^k)$:inventory variance at the k-th stage.

Therefore, desirable systems are as follows;

$$1.0 \geq \mathrm{Amp}(O^1) \geq \mathrm{Amp}(O^2) \geq \mathrm{Amp}(O^3) \ \ldots\ldots\ldots\ldots \geq \mathrm{Amp}(O^K)$$

$$1.0 \geq \mathrm{Amp}(B^0) \geq \mathrm{Amp}(B^1) \geq \mathrm{Amp}(B^2) \ \ldots\ldots\ldots\ldots \geq \mathrm{Amp}(B^{K-1})$$

where, the first stage is a final production or an assembly stage and the K-th stage is a raw material processing stage.

A change for the worse in Amp(O) and Amp(B) is serious, causing increases in lead time, inventory levels and costs.

In machine processing systems of multi-stage production systems, preventing large increases in lead time and inventory is needed to shorten processing and

TABLE 1 Factors Affecting Flexibility in Single Stage Manufacturing Systems

Evaluative Variables / Type of Process	Variety of Parts or Product	LT Minimization	Inventory Minimization	Investment Savings	Cost Savings
Machining Production	·$t_s \to$min	·t_s, t_m $\to$min	·$t_h \to$min	·Depends on equipment	·Labor cost savings realized by automation & robotization ·Depends on investment cost
Lot Production	·$t_s \to$min ·$q \to$min	·t_s, t_m $\to$min ·$q \to$max	·t_s, t_h $\to$min ·$q \to$max	·Depends on equipment	·Labor cost savings realized by automation & robotization ·Depends on investment cost
Assembly Line Production	·Change to multi-model mixed assembly line production system ·Versatile operators ratio $\to$max ·$t_m \to$min ·Number of work stations $\to$min ·Sequencing			·Savings due to over-lapping equipment	·Saving labor and inventory

Notation t_s: set up time

t_m: manufacturing time per piece

t_h: handling and transportation time

q : production lot size

versatile operator ratio = (total number of work stations mastered by erch worker) / {(number of work stations)*(number of workers)} set up time. Consideration of job shop scheduling problems on multi-job and multi-process, group technology(GT), substitute machinery and the ratio of versatile operators is also important in order to try to adapt the system to management's needs.

In the multi-stage lot production systems, the prevention of large increases in lead time and inventory are needed not only to minimize set up time, processing time and feeding and moving out time but is also needed to hold the ordering system to the optimal lot size, for minimizing total lead time, for preventing work congestion between one stage and the immediate preceding stage, and for increasing flexibility by utilizing the ratio of versatile operators and available buffer systems (refs. 15, 16).

In the multi-stage mixed type production systems, consisting of machining processes, lot production processes and multi-model mixed assembly line systems (ref. 17), the available buffer systems and the ratio of versatile operators are needed to increase flexibility. Factors which increase flexibility in multi-stage production and inventory systems are shown in Table 2.

Reducing set up, processing and feeding and moving out time is accomplished by improvements in equipment and computer-aided engineering. The improvement of ordering systems, determination of optimal lot size and the solving of job sequence problems are accomplished by improving and innovating production management technology and computer-aided engineering. The group technology and designing of multi-model mixed assembly lines are developed through improvements and innovations in equipment, production mangement and computer-aided engineering.

Thus, the term "FMS" describes the following:

Flexible Machining Systems - systems which have automatic processing, tooling, loading and unloading, and handling.

Flexible Manufacturing Systems - systems which are concerned with scheduling and consider the demand of immediately succeeding stages based on consideration of the flexible machining systems.

Flexible Management Systems - systems which totally optimize the flexible manufacturing systems with the flexible machining systems.

A COMPARATIVE STUDY ON THE FLEXIBILITY OF TWO TYPES OF ORDERING SYSTEMS

There are two types of production ordering systems. One is the "push type" production ordering system and another is "pull type" production ordering system .

Concept of "push type" and "pull type" production ordering systems

In the "push type" production ordering system, the ordered quantity in each stage is determined by forecasted demand. Forecasted demand is the length of

TABLE 2 Factors Affecting Flexibility in Multi-stage Manufacturing System

Evaluative Variables / Type of Succession Systems	Variety of Parts or Products	LT Minimization	Inventory Minimization	Amp(O) Minimization	Amp(B) Minimization
Machining Processes	GT	• t_s, t_m and $t_h \rightarrow$ min • Job shop scheduling • Substitute machine • Versatile operator			
Lot Production Processes (two stage)		• t_s, t_m and $t_h \rightarrow$ min • Optimizing production lot size in multi-stage $N^* = \begin{cases} [\sqrt{R_1 t_{m_{1,2}} / \sum_{i=1}^{M} t_{s_{i,1}}}] & (3) \\ [\sqrt{R_M t_{m_{M,1}} / \sum_{i=1}^{M} t_{s_{i,2}}}] & (4) \end{cases}$		• Desirable ordering system • Available buffers	
Mixed Assembly Line, Lot Production and Machining Processes		• t_s, t_m and $t_h \rightarrow$ min		• Desirable ordering system • Sequence of multi-model in final assembly line • Available buffers • Versatile operator	

where, N : number of setups

R_i: production quantity of the i-th item per period

$\tau_{i,k}$: production time of the i-th item at the k-th stage

$Amp(O^k) = V(O^k)/V(D)$

$Amp(B^k) = V(B^k)/V(D)$

$V(x)$: Variance of x

(3) $\sum_{i=1}^{M} \tau_{i,2} \leq \sum_{i=1}^{M} \tau_{i,1}$ and

$\tau_{i+1,2} \leq \tau_{i,1}$

(for $i=1,2,\ldots,M-1$)

(4) $\sum_{i=1}^{M} \tau_{i,2} > \sum_{i=1}^{M} \tau_{i,1}$ and

$\tau_{i+1,2} > \tau_{i,1}$

(for $i=1,2,\ldots,M-1$)

cumulated lead time from one stage to the final assembly line, and of feedback information of product or in-process inventory in each stage. In this system, the ordered quantity of each production stage is ordered by a central controller.

Thus, it may also be called a "centralized ordering system". Material flows are controlled just as if they are "pushed out" from the raw materials stage toward the final product stage.

In "pull type" production ordering systems, the ordered quantities in each stage are determined by actual quantities consumed by the immediate downriver stage. Here no central controller is needed. Thus, it may also be called a "decentralized ordering system". Material flows are controlled just as if they were "pulled" into the final product stage from the stages preceeding final production.

General model formulation

Assumptions.

1) Figure 1 and 2 show the schematic diagram of "push type" and "pull type" production ordering systems.
2) The systems consist of K production stages. Each production stage has only one process. Each production process produces M kinds of products.
3) There are two kinds of the inventory stages. The inventory stage $I^{k(i)}$ is the part inventory for the i-th product which has been fabricated by the production stage k. The inventory stage $B^{k(i)}$ is the part inventory for the i-th product which is the on-hand material level for the production stage k. The inventory stage $B^{0(i)}$ is the final product inventory stage of the i-th product.
4) The production lead time for the k-th stage is described as L_p^k. The production quantity ordered to the stage at the end of the T-th period is completed during the $(T+L_p^k)$th period and is stored in the fabricated inventory $I^{k(i)}$ at the end of the $(T+L_p^k)$th period.
5) The handling and transportation lead time from the fabricated inventory $I^{k(i)}$ at the k-th stage to the on-hand inventory $B^{k-1(i)}$ at the (k-1)th stage is described as L_h^k.
6) Back logs are permitted.
7) The raw material inventory supplied for the K-th stage is always sufficient. But at the other stages 1,2,....,K-1, the quantity ordered for production is restricted by the on-hand materials inventory $B^{k(i)}$.
8) The production capacity of each stage is the same as in each planning period
9) Down time occurs at each production stage. During down time, the production stage stops producing.

Notations

L_h^k : handling and transportation lead time at the k-th stage,

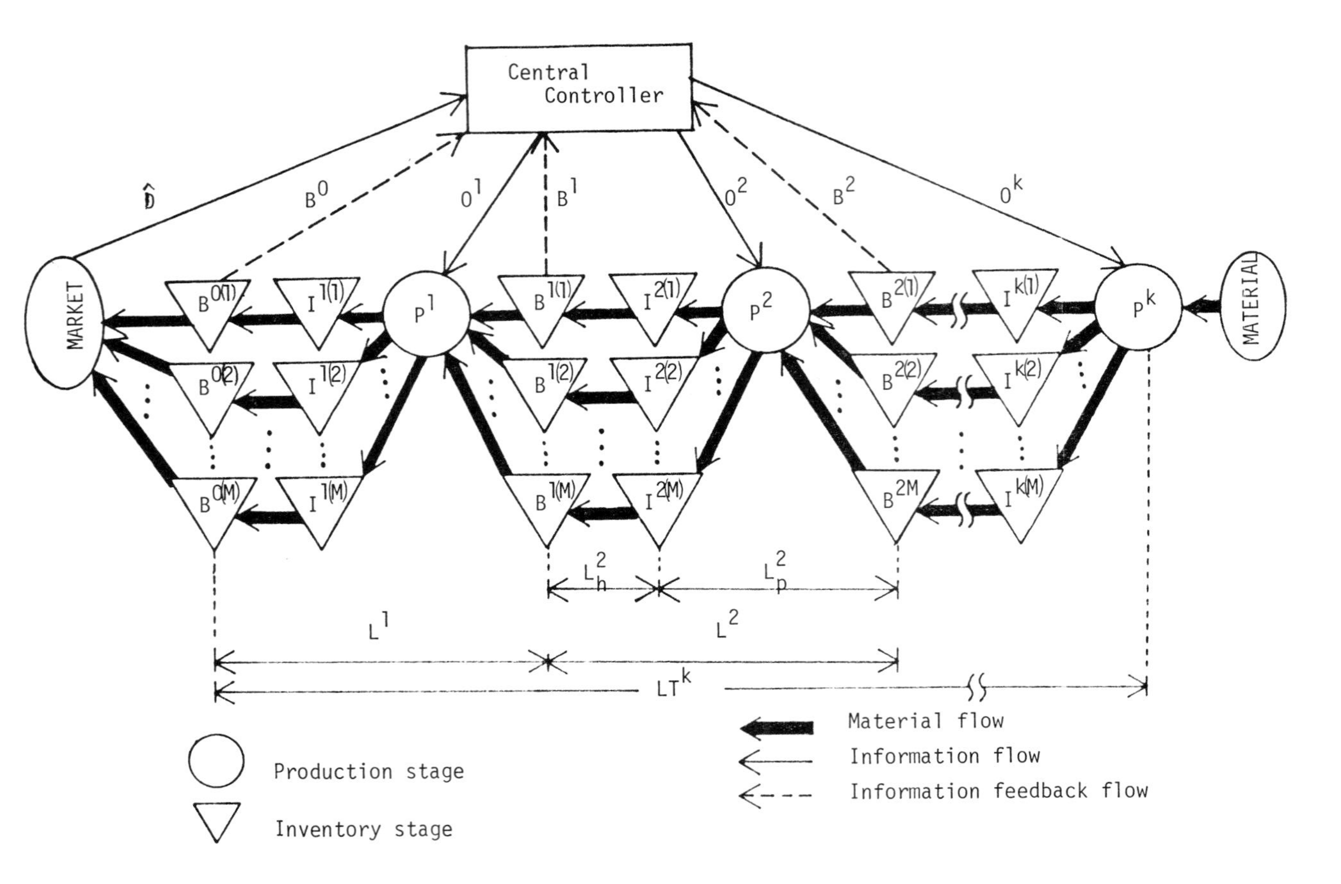

Fig.1. Schematic Diagram of the Push Type Ordering System

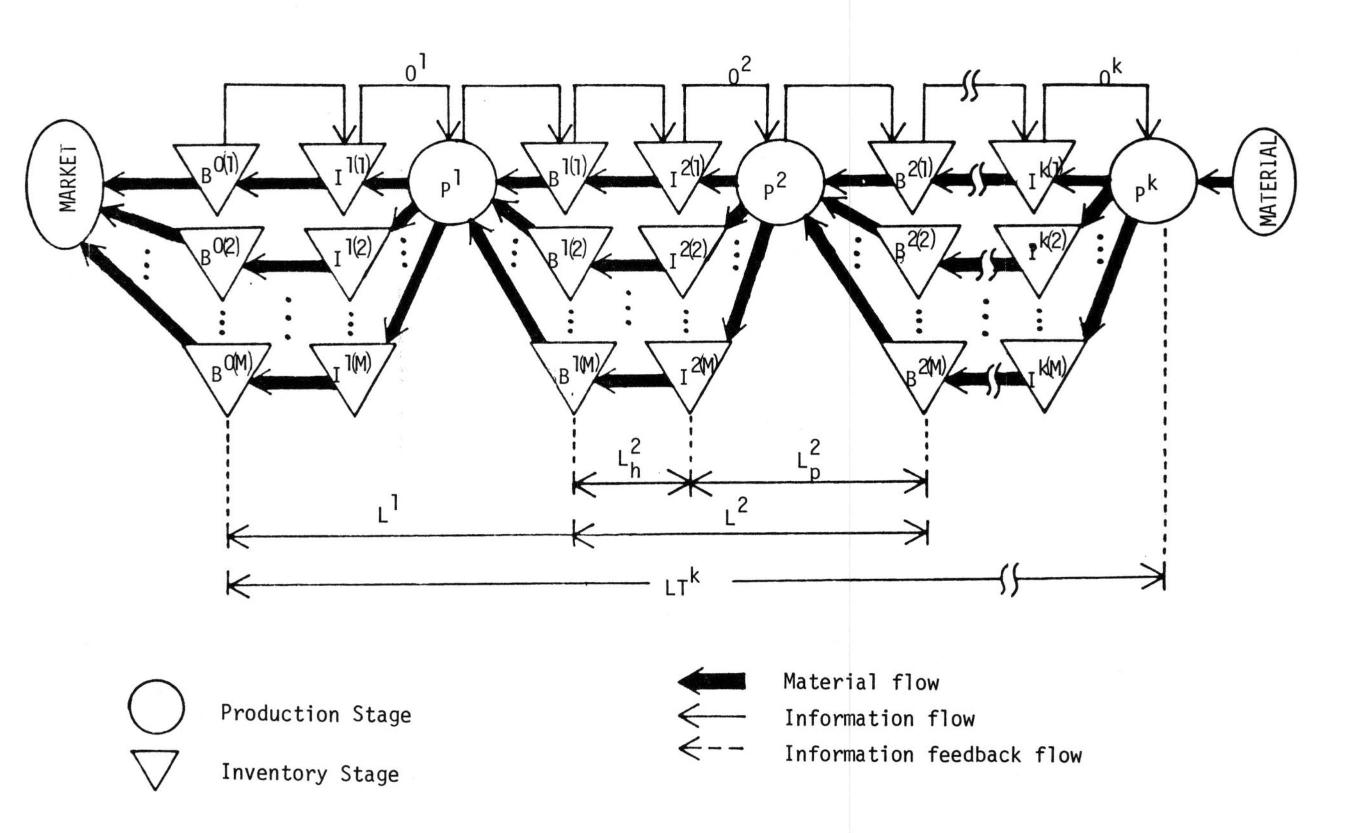

Fig.2. Schematic Diagram of the Pull Type Ordering System

L_p^k : production lead time at the k-th stage,

L^k : the lead time from production starting time at the k-th stage to the handling and transportation completion time at the (k-1)th stage, i.e.

$$L^k = L_h^k + L_p^k \tag{2}$$

LT^k: the accumulated lead time from the final stage to the k-th stage, i.e.

$$LT^k = \sum_{k'=1}^{k} L^{k'} \tag{3}$$

$\hat{D}_{T:T+L}^{(i)}$: the forecasted market demand of the i-th product for the (T+L)th period forecasted at the end of the T-th period,

$D_T^{(i)}$: the actual market demand of the i-th product in the T-th period,

$Q_{T:T+L^k+1}^{k(i)}$: the required quantity of the i-th product for the $(T+L^k+1)$th period at the k-th stage calculated at the end of the T-th period,

$O_{T:T+L^k+1}^{k(i)}$: the ordered quantity which is determined on the basis of the required quantity with restrictions on production capacity and material inventory.

$P_T^{k(i)}$: the actual production quantity of the i-th product during the T-th period in the k-th production stage,

$I_T^{k(i)}$: the inventory of the i-th product fabricated by the k-th production stage at the end of the T-th period,

$S^{k(i)}$: the safety stock of the i-th product at the k-th stage,

C^k: the production capacity at the k-th stage during the T-th stage,

X^k: the down time at the k-th production stage during the T-th period,

$e_{T-L:T}^{(i)}$: the forecasting error of the i-th product for the T-th period forecasted at the end of (T-L)th period, i.e.

$$e_{T-L:T}^{(i)} = \hat{D}_{T-L:T}^{(i)} - D_T^{(i)} \tag{4}$$

A(i,k): the required part quantity of the k-th production stage to manufacture one unit of the i-th product.

System equations

(1) Push type production ordering systems

$$Q_{T:T+L^k+1}^{k(i)} = A(i,k)\cdot\hat{D}_{T:T+LT^k+1}^{(i)} + \sum_{\ell=1}^{L^K} A(i,k)\cdot\hat{D}_{T:T+LT^{k-1}+1-\ell}^{(i)} - \sum_{\ell=1}^{L^k} Q_{T-\ell:T+L^k-\ell+1}^{k(i)} - B_T^{k-1(i)} + S^{k-1(i)}$$

$$\text{(for } k=1,2,3,\ldots\ldots,K \text{ and } i=1,2,3,\ldots\ldots,M) \tag{5}$$

$$O_{T:T+L^k+1}^{k(i)} = \min\{Q_{T:T+L^k+1}^{k(i)},\ C^k\cdot Q_{T:T+L^k+1}^{k(i)} / \sum_{i'=1}^{M} Q_{T:T+L^k+1}^{k(i')},\ (B_T^{k(i)} + P_{T-L_h^k}^{k+1(i)})\cdot A(i,j)/A(i,k+1)\}$$

$$\text{(for } k=1,2,3,\ldots\ldots,K-1 \text{ and } i=1,2,3,\ldots,M) \tag{6}$$

$$P_T^{k(i)} = \min\{O_{T-L_p^k:T+L_h^k+1}^{k(i)},\ (C^k - X_T^k)O_{T-L_p^k:T+L_h^k+1}^{k(i)} / \sum_{i'=1}^{M} O_{T-L_p^k:T+L_h^k+1}^{k(i')}\}$$

(for k=1,2,3,....,K and i=1,2,3,....,M) (7)

$$I_T^{k(i)} = P_T^{k(i)} \quad \text{(for k=1,2,3,....,K and i=1,2,3,....,M)} \tag{8}$$

$$B_T^{k-1(i)} = B_{T-1}^{k-1(i)} + P_{T-L_h^k-1}^{k(i)} - P_T^{k-1(i)} \cdot A(i,k)/A(i,k-1)$$

(for k=2,3,4,...,K and i=1,2,3,...,M) (9)

$$B_T^{0(i)} = B_{T-1}^{0(i)} + P_{T-L_h^k-1}^{1(i)} - D_T^{(i)} \quad \text{(for i=1,2,3,...,M)} \tag{10}$$

(2) Pull type production ordering systems

$$Q_{T:T+L^1+1}^{1(i)} = D_T^{(i)} + (Q_{T-L_p^1,T+L_h^1+1}^{1(i)} - P_T^{1(i)}) \quad \text{(for i=1,2,3,...,M)} \tag{11}$$

$$Q_{T:T+L^k+1}^{k(i)} = P_T^{k-1(i)} \cdot A(i,k)/A(i,k-1) + (Q_{T-L_p^k:T+L_h^k+1}^{k(i)} - P_T^{k(i)})$$

(for k=2,3,4,....,K and i=1,2,3,...,M) (12)

$$O_{T:T+L^k+1}^{k(i)} = \min\{Q_{T:T+L^k+1}^{k(i)}, C^k Q_{T:T+L^k+1}^{k(i)} / \sum_{i'=1}^{M} Q_{T:T+L^k+1}^{k(i')}, (B_T^{k(i)} + P_{T-L_h^k}^{k+1(i)}) A(i,k)/A(i,k+1)\}$$

(for k=1,2,3,...,K-1 and i=1,2,3, ...,M) (13)

$$P_T^{k(i)} = \min\{O_{T-L_p^k:T+L_h^k+1}^{k(i)}, (C^k - X_T^k) O_{T-L_p^k:T+L_p^k+1}^{k(i)} / \sum_{i'=1}^{M} Q_{T-L_p^k:T+L_h^k+1}^{k(i')}\}$$

(for k=1,2,3,....,K and i=1,2,3,...,M) (14)

$$I_T^{k(i)} = I_{T-1}^{k(i)} + P_T^{k(i)} - P_{T-1}^{k-1(i)} \cdot A(i,k)/A(i,k-1)$$

(for k=1,2,3,...,K and i=1,2,3,...,M) (15)

$$B_T^{0(i)} = B_{T-1}^{0(i)} + P_{T-L_h^1-1}^{(i)} - D_T^{(i)} \quad \text{(for i=1,2,3,....,M)} \tag{16}$$

$$B_T^{k-1(i)} = B_{T-1}^{k-1(i)} + P_{T-L_h^1-1}^{k-1(i)} \cdot A(i,k)/A(i,k-1) - P_T^{k-1(i)} A(i,k)/A(i,k-1)$$

(for k=2,3,4,....,K and i=1,2,3,....,M) (17)

Basic model analysis

Some results of an analysis of a basic model which has no restrictions on production capacity, inventory and down time are presented here to clarify the characteristics of the two types of production ordering systems.

Basic model assumptions. Basic models of two types of production ordering systems are simplified here. The parameters of the systems are as follows;

1) K=9
2) M=1 and A(1,k)=1 for k=1,2,3,....,9

3) $L_p^k = 1$ and $L_h^k = 0$ for $k=1,2,3,\ldots,9$

4) There are irregular variations in the demand time series.

5) The forecasted demand is expressed as follows;

$$\hat{D}_{T:T+L} = \overline{D} + e_{T:T+L} \tag{18}$$

$$e_{T:T+L} = a \cdot L \cdot u \tag{19}$$

where, a: the degree of the forecasted error depends on the length of lead time in the forecasting period.

u: the unit disturbance depends on the normal distribution $N(0,1^2)$.

6) The production capacity at each stage is always sufficient.

7) The amount of safety stock at each stage is always sufficient.

8) Down time at each stage does not occur.

9) The performance measurement of the systems is defined as follows;

$$\begin{aligned} Amp(O^k) &= V(O^k)/V(D) \\ Amp(B^k) &= V(B^k)/V(D) \end{aligned} \tag{1}$$

The "amplifications" in production ordering and inventory quantity. Fig.3 and Fig.4 show the "amplifications" in production ordering and inventory quantities on the two types of production ordering systems.

Results. This analysis has the following conclusions:

1) In the push type production ordering system, the "amplifications" in production ordering and inventory quantities at each stage are more diffused in the stages further to final production. A larger forecasting error is likely the greater "amplifications" in production ordering and inventory quantities at the stages. Therefore, a control parameter is needed in the push type production ordering system to prevent these "amplifications".

2) In the pull type production ordering system, there is no "amplification" in the production ordering and inventory quantities. The production ordering systems without "amplifications" are effective in stabilizing the multi-stage systems and in increasing the flexibility and productivity of the systems. However, it is necessary that inventory levels are decreased at each stage in this production ordering system.

CONCLUSIONS

In recent years, many flexible manufacturing systems have been developed throughout the developed countries. It is important to design and operate production systems which adapt to severe changes in the market and in technology and to increase productivity.

In this paper, the evaluative variables of flexibility are proposed and factors improving the flexibility of single stage production systems which have different characteristics are discussed.

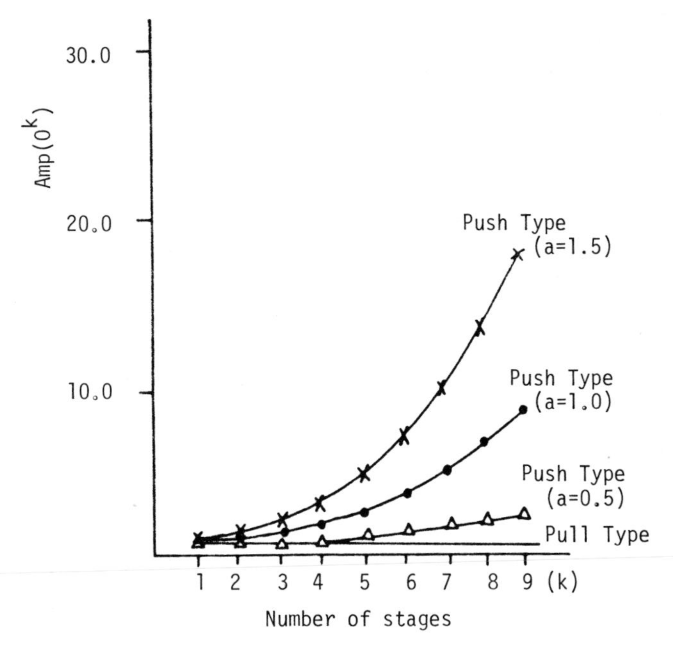

Fig.3. "Amplifications" of the Ordered Quantity at Each Stage

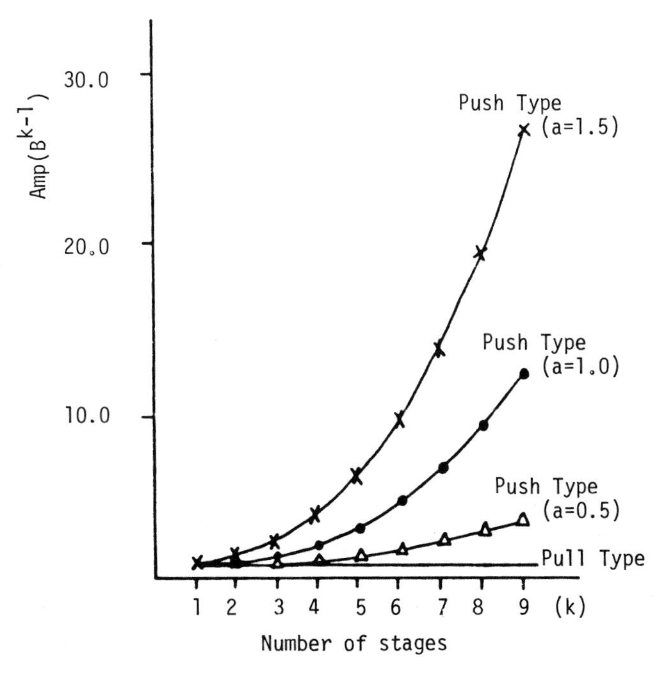

Fig.4. "Amplifications" of the Inventory Quantity at Each Stage

Currently, most company production systems consist of multi-stage mixed type systems. This paper also points out that the "amplification" in multi-stage production systems is one of the important factors affecting the system's flexibility and productivity. The paper also discusses some ways of reducing this "amplification" by utilizing available production ordering systems.

The multi-stage production ordering system models discussed here are simplified, and neglect the characteristics of each single production system which consist of total multi-stage production systems. The models also neglect transportation and material flow parameters.

The flexible manufacturing system of today is merely treated as a stage in a succession of production stages with different characteristics; however, it should be considered as a way to increase the flexibility and productivity of the system as a whole.

REFERENCES

1 J.Hartley, FMS at work, North-Holland, Amsterdam, 1984, 286 pp.
2 K.Takeda, S.Shimoyashiro and N.Tsuchiya, Technology and Practice of Signal System, Proceedings of 7th ICPR in Windsor, 1983, 563-566.
3 A.Masuyama, Idea and Practice of Flexible Manufacturing System of Toyota, Proceedings of 7th ICPR in Windsor, 1983, 584-590.
4 N.Mizoguchi, Flexible Manufacturing System for Photo-conductors, Proceedings of 7th ICPR in Windsor, 1983, 591-597.
5 J.Parnaby, Concept of a Manufacturing System, International Journal of Production Research, 17, 2 (1979) 123 - 135.
6 D.M.Zelenovic, Flexibility - A Condition for Effective Production Systems, International Journal of Production Research, 20, 3 (1980) 319- 337
7 R.E.Young, Software Control Strategies for Use in Implementing Flexible Manufacturing Systems, Industrial Engineering, 13, 11 (1981) 88-96
8 K.E.Stecke, Loading and Control Policy for Flexible Manufacturing System, International Journal of Production Research, 19, 5 (1981) 481-490
9 J.A.Bazacott, The Fundamental Principles of Flexibility in Manufacturing Systems, Proceedings of the 1st International Conference on Flexible Manufacturing Systems, Brighton, UK, October (1982) 13-22
10 D.J.McBean, Concepts of FMS, Proceedings of the 1st International Conference on Flexible Manufacturing Systems, Brighton, UK, October (1982) 497-513
11 S.M.Gustavsson, Flexibility and Productivity in Complex Production Process, Proceedings of International Conference on Productivity and Quality Improvement, Tokyo, Japan, October (1982) B-3-1
12 P.G.Ranky, The Design and Operation of FMS - Flexible Manufacturing Systems -, North-Holland, Amsterdam, 1983, 348 pp.
13 R.Muramatsu, H.Miyazaki and Y.Tanaka, Effective Production Systems which harmonized Worker's Desires with Company Needs, International Journal of Production Research, 20, 3 (1982) 297 -309
14 T.Tabe, R.Muramatsu and Y.Tanaka, Analysis of Production Ordering Quantities and Inventory Variations in Multi-Stage Production Ordering Systems, International Journal of Production Research, 18, 2 (1980) 245-257
15 Y.Tanaka and R.Muramatsu, A Study of the Design of a Lot Production System, International Journal of Production Research, 15, 6 (1977) 565 -581
16 S.Kubokawa and M.Sosiroda, Scheduling Procedures for Special Order Items in a Two-stage Lot Production System, International Journal of Production Research, 18, 1 (1980) 43 - 56

17 Y.Tanaka and R.Muramatsu, An Analysis of Dynamic Characteristics of Two-stage Mixed Products Line Production Process and Lot Production Process Model, International Journal of Production Research, 20, 5 (1982) 629- 641

FLEXIBLE MANUFACTURING SYSTEM FOR PHOTO-CONDUCTORS

Noboru Mizoguchi

Fuji Electric Company Ltd., Japan

1. Outline of Matsumoto Factory of Fuji Electlic Co., Ltd.

The Matsumoto Factory manufactures semiconductor products such as photo-conductors, silicon semiconductors, amorphous silicon solar cells, etc., and electro-mechanical products, such as watthourmeters, semiconductor applied equipments, data processors, peripheral terminal devices for information processing etc.

It has about 1,700 employees and is boasting of annual output of about 40 billion yen.

2. Outline of FMS for photo-conductors

2-1. Photo-conductor

The photo-conductor is used as a sensitive element of dry type copying machines.

The sensitivity and size perfectly match the construction of the copying machines.

The user is a large number of copying machine manufacturers.

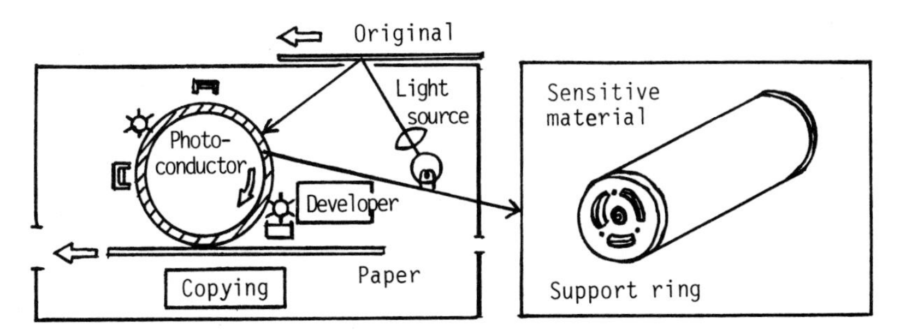

Fig. 1 Construction of copying machine

Fig. 2 Configuration of photo-conductor

2-2. Requirments of customers for photo-conductor

(a) Sharp image
(b) Quick copying speed
(c) Long product life cycle
(d) Confomity to model change
(e) Wide variatinal product specifications

2-3. Situation of photo-conductor production

(a) Number of basic model varies 32 types.
(b) The number of changes in the orders received at once is 32 in monthly average (50 in maximum and 15 in minimum).
(c) Maximum fluctuation of demands among individual basic models are about five times as that of total demands.

2-4. Basic concept of flexible factory automation system (F.F.A.S.)
Refer to Fig. 3

The flexible factory automation system in Fuji Electric Co., Ltd. is an integrated system dealing with business activities, designing, manufacturing, testing and shipping to realize high-quality, low-cost products and to answer a variety of requirements from users, such as diversification, small order lot size and quick delivery.

The structure of the system is as follows.

The system is combined with sales and engineering system and the factory management systems.

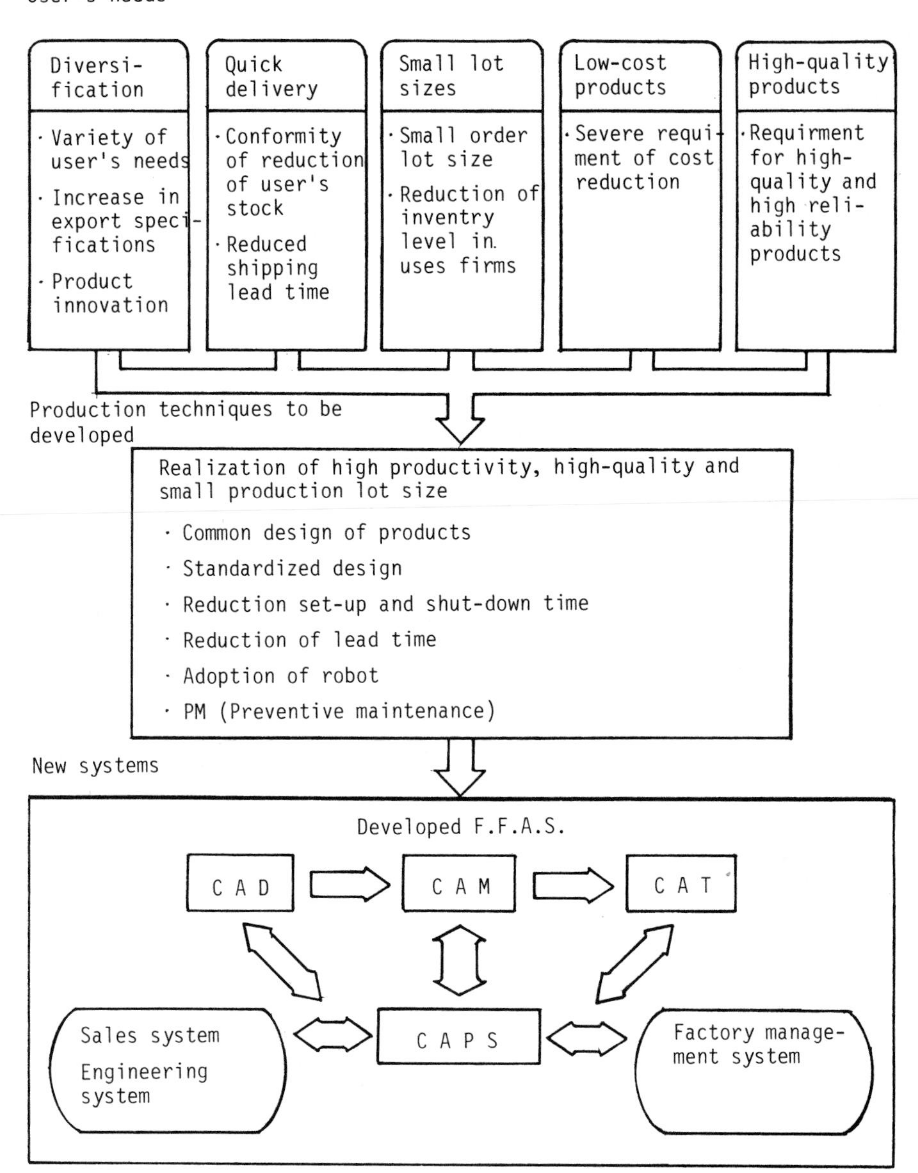

Fig. 3 Basic concept of flexible factory automation system (F.F.A.S.)

3. Objectives of development of F.F.A.S. for photo-conductors

3-1. Problems in old system

Number of basic modles of photo-conductor is increasing, due to the diversification of user's needs.

In this case, the following problems about management have arised in the manufacturing system.

(a) Relationship between characteristics of products required by users and condition of the manufacturing process is inadequate.

(b) Increase of defect percentage due to workers required, careless setting and supervision of manufacturing conditions.

(c) Increase of cost due to the diversification of products.

(d) Erroneous capacity load plan due to diversification of products.

(e) Diversification is difficult because it takes much arrangement time.

(f) Appropriate countermeasure with user's request after planning a capacity load plan is difficult.

3-2. The major point of design policy and operation of new system

To cope with those problems, the development of F.F.A.S. was begun in 1981.

The major points of production-research to realize are shown below in Table 1.

TABLE 1 The major point of the development of F.F.A.S.

Purpose	Major item to be improved by management needs
High quality	1) Prevention of artifical variations of manufacturing process control. 2) Development of sensor electronics. 3) Stabilization of environment involving temperature, humidity and dust which influences the characteristics of products. 4) Reduction of down time in processings which influences the variety of products and maintenance. 5) Supply of reliable quality evaluation data required of users. 6) Acquisition of market secular change data, company evaluation data and other data of manufacturing process conditions, sensitive material, etc., and application of thease data for improvement of quality of products.
Low-cost products	1) In order to save labor cost, automated system for manufacturing process and transport. 2) Robotization for assembly and package operations.
Diversified, small product lot sizes	1) High speed setting of manufacturing process conditions. 2) Robotization for preparation of tool setting and reduction of set-up and shut-down time. 3) Development of computer aided schedules.
Quick delivery	1) Reducation of dead time and manufacturing lead time by integrated line system for quick compliance to change in production schedules. 2) Reduction of time for simplification of capacity load plan.

4. Features of new system

The newly developed system has the following characterisitics.

4-1. High quality

(a) The adoption of automated system in the composite line for machining → washing and heat treatment → coating → inspection → assembly → package, stabilizes environmental condition and decreased down time.

(b) Manufacturing process condition of each line is controlled by automated system, elimination of artificial variations of the quality of products.

(c) Automatic material loading system for dimension inspection and rejection.

(d) In-process control by using electronic sensor for film coating.

(e) Computer controlled hierarchy processing system for machining data, measurement data, achievement data and quality data.

(f) Automated system for preparation and automated supply system for inspection data by users.

(g) In-process control of inspection condition.

(h) Bar code is attached to individual products in the stage of processing to control the manufacturing process condition and quality data of products for compliance to the market evaluation data.
The data are analyzed, as necessary, and feed back to the manufacturing process condition control.

4-2. Low-cost products

(a) The chuck type robot which is developed by factory and is free from deformation, is used for loading and unloading with in and between processes.
(b) Development of assembly and package systems in which robots works automatically.
(c) Improvement of utilization ratio by automatic data collection and arrangement of operating data of line facilities.

4-3. Diversified, small user's lot sizes products

(a) Automatic setting of manufacturing condition by computer.
(b) Automatic exchange of tools by computer.
(c) String of one shift materials and supply in sequence according to production schedule.
(d) Selection and setting of film coating condition (cassette system) by computer.
(e) Automatic setting of inspection condition by computer. (Automatic program reference by types)
(f) Automatic setting of washing condition of ultrasonic washing machine by computer.
(g) Automated system for loading and unloading, and reduction of the time.
(h) Minimization of in-process inventry between processes by automatic transport and supply systems.

4-4. Quick delivery

(a) Collection and application of data of capacity load plan, manufacturing, achievement, receipts of orders, and quality.

5. Outline of F.F.A.S. for photo-conductor

Fig. 4 Shows the line configuration
Fig. 5 Shows the data system diagram for quality and manufacturing control.

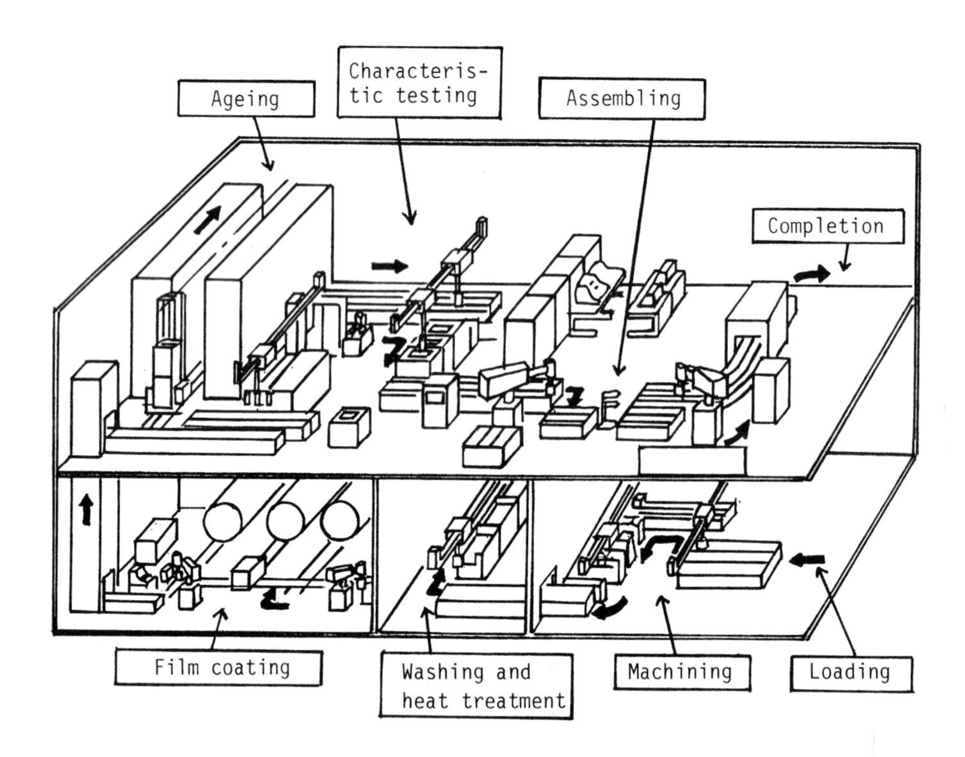

This system consists of a total 30 machines for machining, washing, film coating, inspection and assembly plus 16 robots and as manual conveyors.

This system is designed to perform all processings from loading of materials to completion of products by means of a mini-computer and micro-computer hierarchy system.

Fig. 4 Line configuration diagram

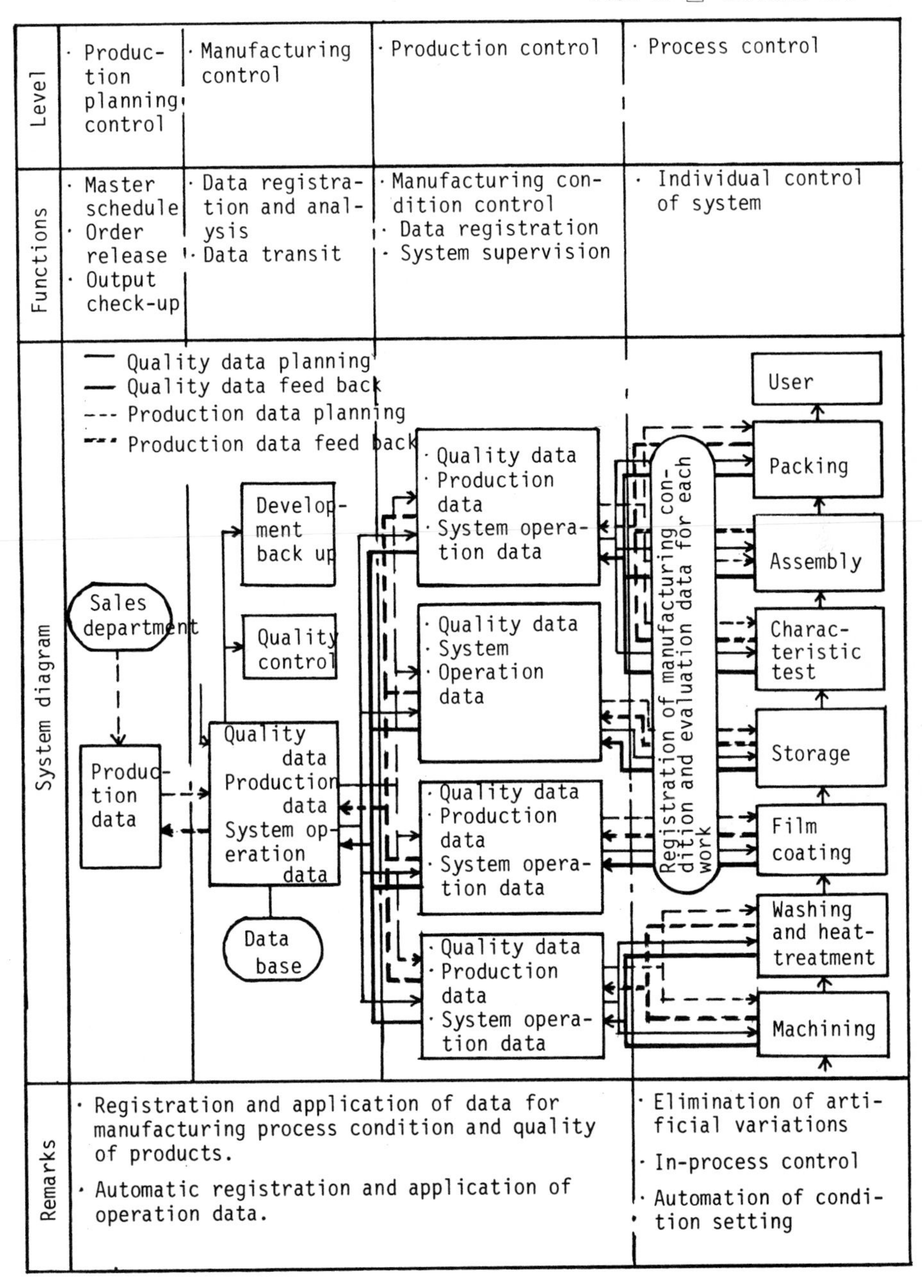

Fig. 5 Data system diagram for quality manufacturing control and information

6. Progress of system development

Time schedule of F.F.A.S. development is shown in table 2.

TABLE 2 Time schedule of F.F.A.S. development

Development item	Schedule			
	1980	1981	1982	1983
6-1. Standardization of processings, and registration and application of data				
(a) Improvement of manufacturing process condition for standardization		←	→	
(b) Standardization of environment (temperature, humidity, deposits of dust) and reduction and standardization of dead time in process.		←	→	
(c) Registration of data for quality and process condition, and development of application system and development of application system for quality and process condition data in the stage of product development.		←	—	→
6-2. Development of automation technology of assembly and transportation				
(a) Development of robot for automated assembly technology, for automated packing technology and for loading and unloading and development of facility fault detecting system.		←	—	→

6-3. Development of automated arrangment technology

(a) Development of automated tool exchange system for NC machines

(b) Development of automated setting program for film coating condition, etc.

(c) Development of automated setting program for inspection condition

6-4. Development of production control system

(a) Development of software for controlling schedules and achievement based on demands and process data.

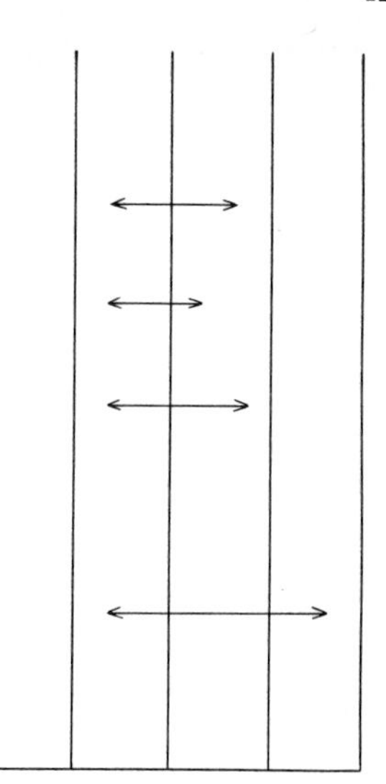

7. Achievement

The results of F.F.A.S. in Fuji Electric Co., Ltd. are shown in figure 6.1 - 6.6.

The broken line shows the achievement of the new system, and the solid line shows the old system.

Each figure is shown differences between the two systems.

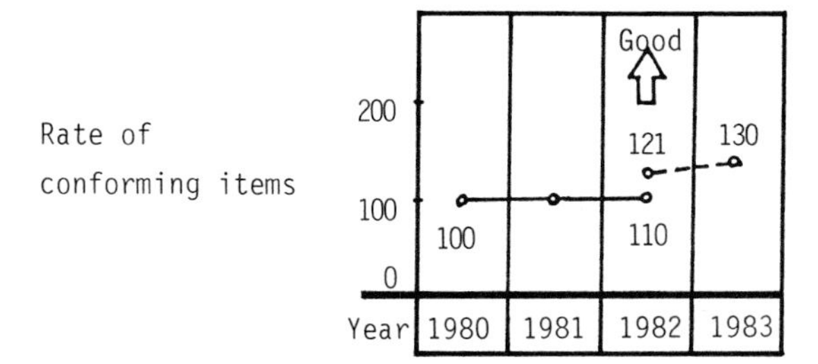

Rate of confroming items:

$$\frac{\text{NO. of finished products}}{\text{NO. of materials loaded}}$$

(1980=100)

Fig. 6-1 Progress of rate of conforming items

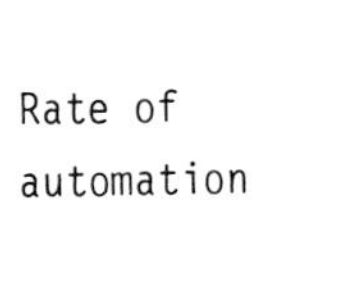

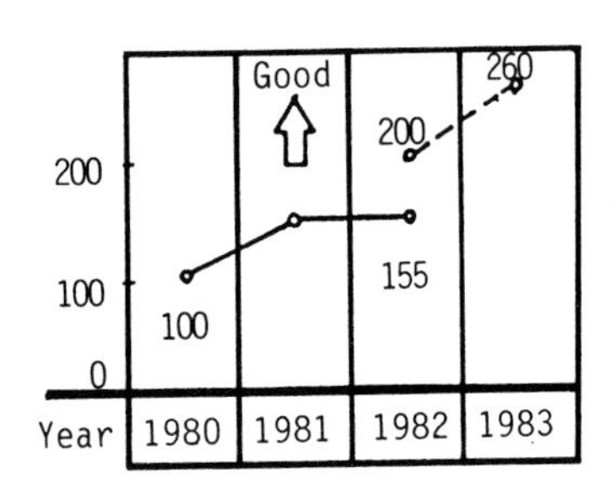

Rate of automation:

$$\frac{\text{Automation (NO. of processes + NO. of items processed)}}{\text{Whole process (NO. of processes + NO. of items processed)}}$$

(1980=100)

Fig. 6-2 Rate of automation

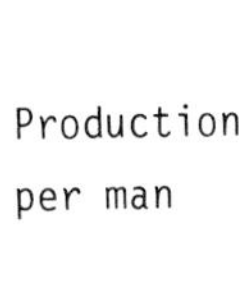

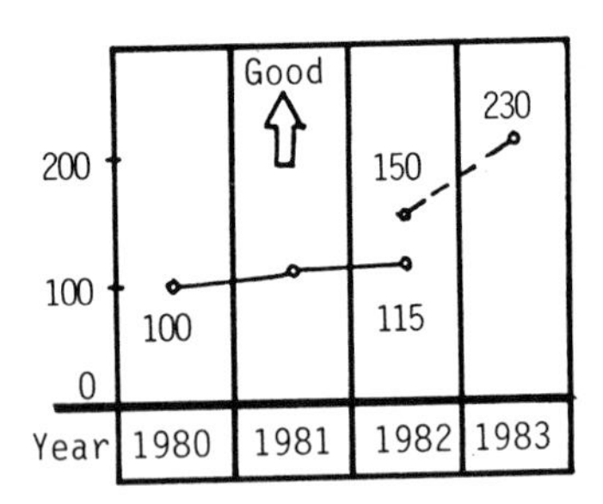

Production per man:

$$\frac{\text{Production per month}}{\text{Number of men per month}}$$

(1980=100)

Fig. 6-3 Production per man

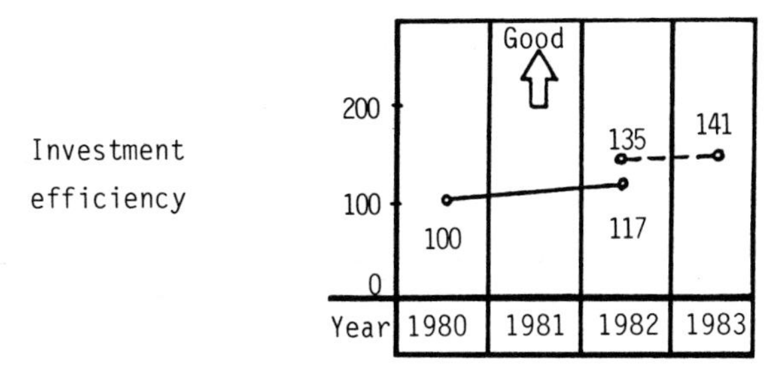

Investment efficiency:

$\frac{\text{Old system}}{\text{New system}}$ of unit cost $\frac{\text{(Fixed cost + Floating cost)}}{\text{Production}}$

(1980=100)

Fig. 6-4 Investment efficiency

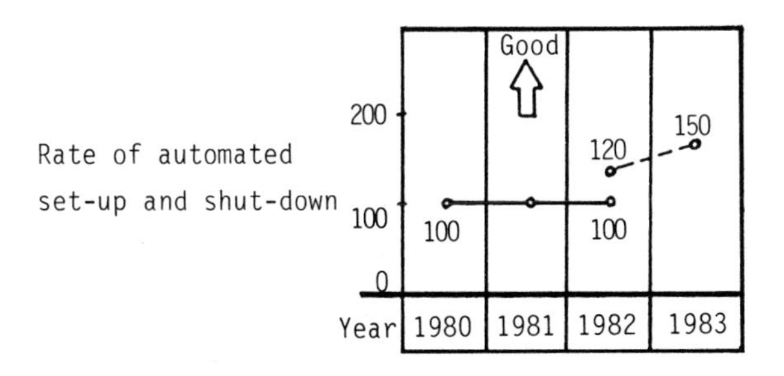

Rate of automated set-up and shut-down:
Ranking of arrangment time in whole line
(OH:100%, less than 3H:95%, etc.)

(1980=100)

Fig. 6-5 Rate of automated set-up and shut-down

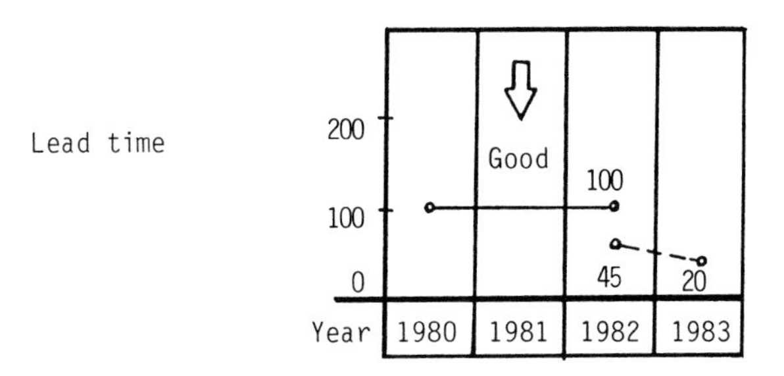

Lead time:
Time required to complete products after loading materials at initial process

(1980=100)

Fig. 6-6 Average lead time

8. Conclusion

The diversified user's needs and the increase of the number of basic models in these days have made it difficult for Fuji Electric Co., Ltd. to respond to the improvement in quality and timely production with conventional automation systems.

To cope with such a problem, Fuji Electric Co., Ltd. has adopted automation systems in each process in order to improve the quality of photoconductor and the productivity and recently completed F.F.A.S. line perfectly meets the management needs, and this new system improved the quality of products and offered high productivity.

Besides the above-mentioned system, Fuji Electric Co., Ltd. is also tackling with the following projects.

(a) By standardization of technology of design and production, the target of further improvement of product quality and productivity will be fit.

(b) The flexible management system, which meets user's needs will be developed.

DESIGN OF PRODUCTION SYSTEMS FOR HIGH FLEXIBILITY AND HIGH CAPITAL UTILIZATION

Jan Lindér
Assistant Professor
Department of Industrial Management
Chalmers University of Technology
S-416 96 Göteborg
Sweden

Abstract

The conditions for efficient production systems are changing. There is an increasing emphasis on fast response to market requirements, at the same time as the pressure for cost reduction forces us to utilize our production factors in a more efficient way.

This paper gives guidelines on how to analyze and design a production system under such conditions.

In the production system in question, manufacturing telephone exchanges, balancing capital cost against disruption cost was the foremost objective. The design of the production system is subject to three demands on the system.

1. Short throughput time
2. Ability to adapt to changes in the environment i.e. flexibility
3. Fast and reliable information systems

The research results show that this can be obtained by using relatively independent production cells as planning points. It is very important to get a good fit between the capacity of the production cells and the design of the information systems, particularly the planning system. Otherwise problems like idle time and delivery delays will occur and this creates a need for security. This is typical for many Swedish companies and results in a lot of material being tied up in the process. To get this fit, a product-oriented division of authority has been realized. To each such area experts on production planning, internal finance and preproduction engineering were delegated. It is also very important to pay attention to the production cells as social systems. The function as a social system might be of crucial importance, because it will involve more complex rôles leading to a need for more advanced regulation.

Introduction

The designing of production systems according to new principles has been closely observed in Sweden for a long time. Evaluations of trials have been made by the Confederation of Swedish Employers (SAF, 1976) and the Metal Workers' Union (Metall-industriarbetarförbundet, 1981), among others. In some cases the trials have also been the object of systematic analyses (e.g. Karlsson, 1979). The objective has mostly been to increase efficiency while taking in consideration the needs of the individual, by balancing technology and work organization.

The production systems have hitherto been built in a relatively stable environment. When the demands on the companies change, the conditions for production system design also change. In Sweden we have had an increasingly marked emphasis on adaptation to market needs. This has led to both a larger number of model options and increased order-oriented production. At the same time the demands for capital rationalization have also increased while technological development of the products has tended to shorten their life-cycle. These tendencies have been felt most tangibly in the electronics industry. In this paper, the conditions for designing production systems, in this type of environment, will be illustrated more closely. The material has developed out of a research project around this question and is largely based on the evaluation of a system in an electronics company in Sweden (Lindér 1980, 1981).

The production system

The products in this case are telephone exchanges for use in companies and institutions. The production process follows the following fundamental steps.

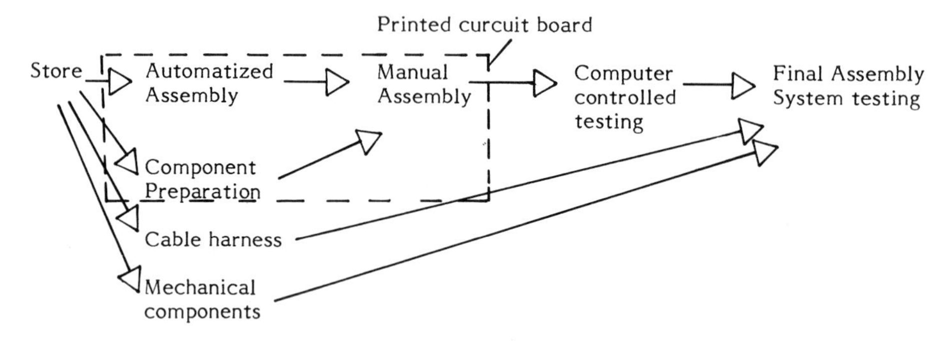

Figure 1 Production Scheme

The solutions discussed later in this paper refer to manual assembly of printed boards, cable harness and mechanical components.

The manufacturing process has distinguishing chacteristics as follows:

- o Rapid technological development gives the products a relatively short life-expectancy. The products are also being constantly developed, leading to design changes.
- o Emphasis on order-oriented production means that specific customer requirements can come close to a promised delivery date and also leads to fluctuation in demand.
- o The supply of electronics components varies and stock-outs are common.
- o The volume means we can classify this as a short-run serial production system.

- o Production is in batches ranging from 25 to 1100 units.
- o The products are relatively complex systems comprising a large number of components.

An analysis of the different elements of the production cost showed wages to account for 13% and materials for 47%. The direct labor cost is as seen relatively small in comparison with the materials cost.

When designing the production system it is important to take into consideration the special characteristics demanded. After analysis the following demands were formulated for the system.

1. A short, reliable throughput time. It is of vital importance to have short throughput times when costumer demands are foremost. The throughput time also is proportionate to inventory investment.
2. Flexibility. Is a term normally used to describe a capacity for adaptation to variation in some way. Here, it refers to the following: 1) Adaption to design change. 2) Adaption to volume change. 3) Adaption to dispatch list change. 4) Adaption to temporary capacity change.
3. Rapid, reliable information systems. A decision-making base is necessary if demands 1. and 2. are to be met. The demands for rapid information are dependent on the length of the planning cycle, planning frequency and disruption frequency.

To evaluate the effects of various solutions, an adapted analytical method was worked out within the project, based on Burbidge (1981), Engström and Karlsson (1981) and Eckerström and Södahl (1981).

The elements of the analytical method are briefly as follows:

1. Total process time. This is proportionate to the labor cost and includes direct labor, set-up time, checking and adjustment time, instruction and foreman's time. Factors influencing total process time are staff motivation, supplied materials quality, set-up time, handling and transport time, balancing losses, skill and instruction time.

2. Queuing times for materials. This is proportionate to the inventory investment for work in process. Factors bearing on this are the need for buffers before operations, the number of operations at the planning point, the need for handling and transport, batch size and the need for a buffer for a given service level for a following link in the chain.

3. Plant costs. This includes the cost of machines, equipment and work area.
4. OH costs. This is proportionate to the share of common resources for planning, management and maintenance used for the system.

System design

The basic element in the design is the planning point (a control point needing one directive for all operations to be carried out). The objective is to integrate as many operations as possible into this point. The problem is difficulty in balancing work-force productivity and capital investment. By reducing throughput time the need for inventory investment is reduced.

Short throughput time ⟶ Reduced inventory investment

One way of reducing throughput time is to increase the frequency of directives, a method tried successfully in Japan (Shingo, 1981). In most cases this leads to increased set-up time cost.

Short throughput time ⟶ Increased frequency of directives ⟶ Increased set-up time cost

At the same time a rapid adaptability is achieved with shorter throughput times.

Short throughput time ⟶ Rapid response to customer demands

This presupposes that it is not more profitable to increase the finished goods inventory. To this must be added that a shorter throughput time reduces the risk of obsolescence and makes it easier to introduce design change at a later date.

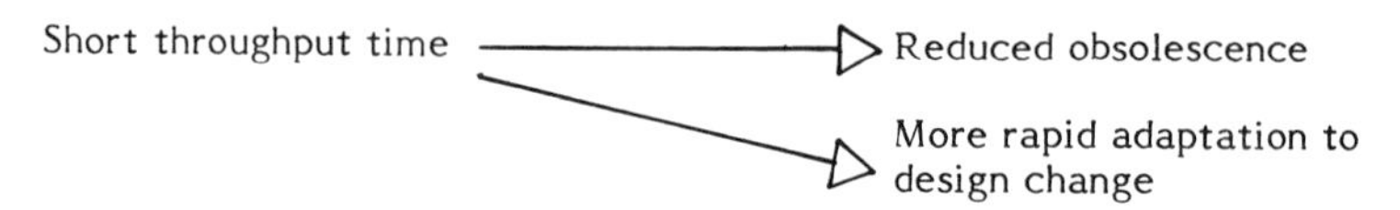

Regarding the company as a whole, disruption sensitivity is reduced. We can more readily adapt to new conditions (the benefit is referable to a rapid response to customer demand) but through our pressured flow we increase disruption sensitively internally. This generates demands for flexibility which we can meet by flexibility in equipment and work organization, and our means could be overcapacity in equipment and investment in training. We then get the following diagram of benefit and costs.

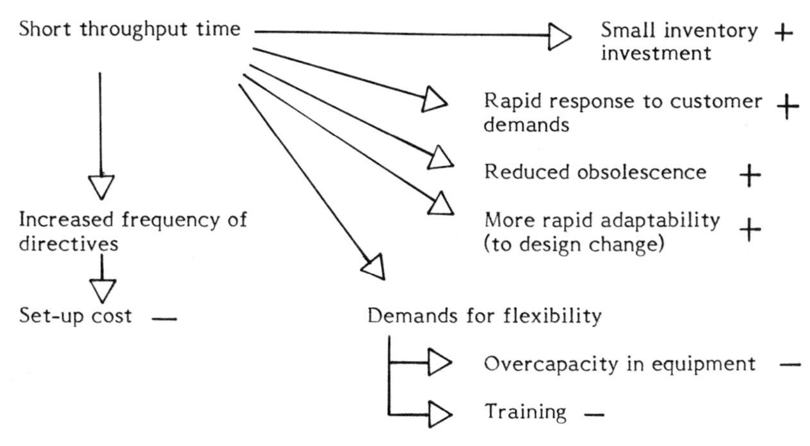

Figure 2 Distribution of benefits and costs

Notice that the set-up cost includes both machine set-up time and the cost for the running-in loss (individual set-up cost).

The diagram below illustrates the running-in time distribution for printed board assembly.

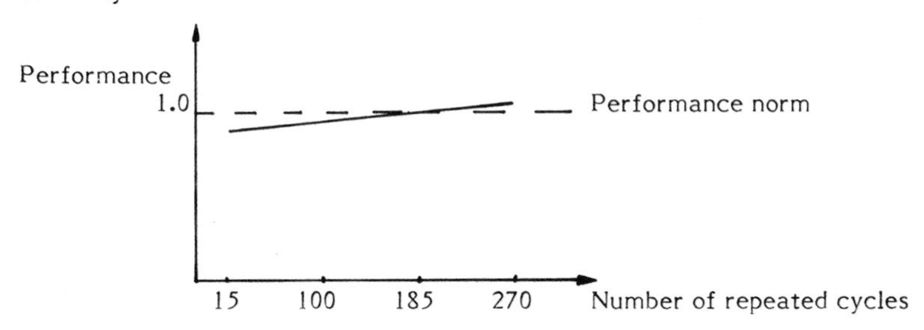

Figure 3 Running-in
(a selection comprising 458 observations over 14 weeks)

Various factors must be considered in the choice of planning point. In this case the analysis showed that a choice of relatively independent production cells as the lowest administrative units would be effective for this system.

Important considerations when establishing a production cell are the part it plays as a social system and its rôle from the point of view of both production and administration.

From the production point of view, the following variables must be taken into consideration:

- number of steps in the process
- set-up cost per step
- $\frac{\text{possible operator density per step}}{\text{operator density per step}}$
- buffer size and position
- common equipment usage
- batch size

The traditional way of dealing with disruptions is with buffers. This possibility diminishes simultaneously as the cost of materials increases due to inventory investment. We must instead increase internal flexibility. Situations to be met are absence (sickness, leave of absence and disruptions in demand and delivery. From a social point of view we must deside what work-rôles the cell is to have and its degree of freedom. This is closely connected to the technical structure which forms the basis for the work-rôles, and to the administrative function of the cell.

As the absorbtion of disruption is transferred from buffers to "internal flexibility" more rapid decision-making is called for, and one way of achieving this end is to make the decisions close to the source of disruption, and relatively independent cells can thereby be effective.

Three conditions must be met for the effective exploitation of such a cell.

1. Assessment norms must be set for the cell's production process allowing clear objectives and result assessment. These comprise time available, quality norms, and norms adapted for the proposes production, (in our case: conditions and clearly defined points of assessment of inventory investment must be fixed).
2. The cell must act as planning point.
   This involves responsibility for:
   - Setting-up a dispatch list within the length of the planning cycle time.
   - Assignment of individuals to tasks, taking into consideration the needs at each operation, i.e responsibility for balancing.
   - To meet the responsibility for for balancing a certain capacity control must exist within this frame of responsibility. Here it is a case of over-time decisions and the right to time off.

- Ordering of material according to the production plan and report obligation for activities in the production process, especially when the product moves on to the next stage.

The (production) cell must assume social responsibility for its members to obviate any undesired effects.

3. The result must be assessed at the planning point and fed-back to it. It is of great importance that the people producing together see a connection between work put in and incentives obtained. This is time both for material incentives and evaluations such as quality control results.

As emphasized earlier, this type of production system demands rapid information processing. In the case in question it was elected to resolve this demand with a result-orienting principle.

Production planning (Ppl), parts of pre-production engineering (Ppr), and internal finance (If) were linked to each result unit and the following diagram shows a schematic representation of this hierarchy.

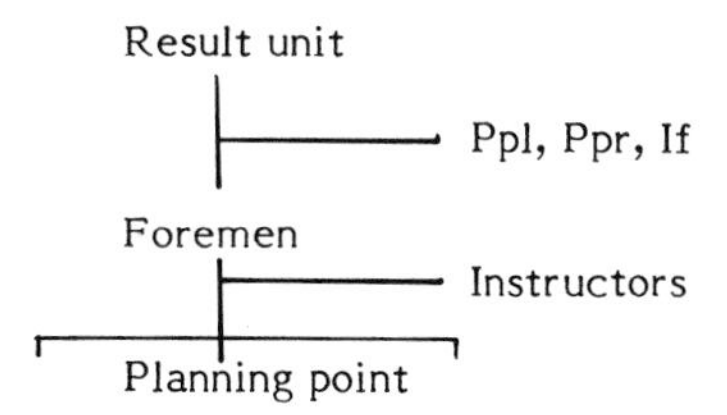

Figure 4 Hierarchy

The introduction of this system gave undesirable effects. It became difficult to maintain the planning process without disruption. A careful analysis showed a distorted contact frequency for the planning function unit.

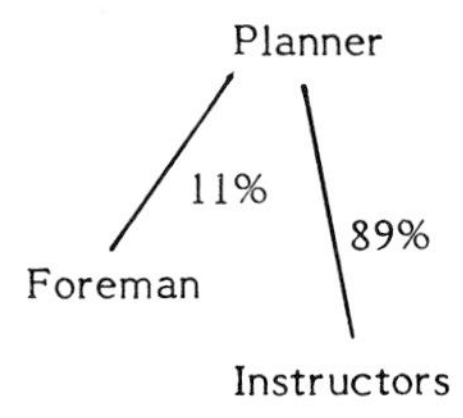

Figure 5 Relative Contact Frequency

The reason for this distribution was that the foreman did not have suitable instruments to measure the result. The lack of adequate information from the foreman forced the planner to contact the production directly via the instructors. This had a double effect: information passed the foreman by and the planner was overloaded.

We see that as greatly reduced buffers impose new demands on the production cell the foreman must also meet new demands. He must have rapid, precise information systems to fulfill his rôle in the planning process.

To solve the problem, an adapted follow-up system was introduced and contact frequency altered as follows:

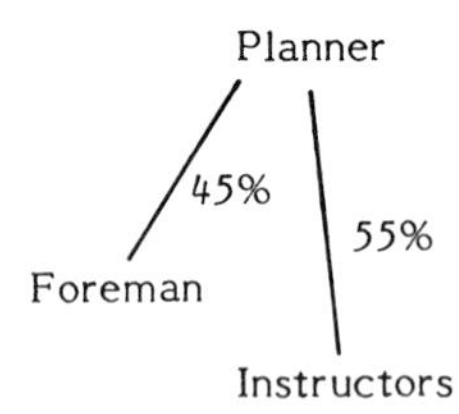

Figure 6 Relative Contact Frequency

An important lesson to be learned from this is that fit is necessary between the capacity of the planning point and the design of the control system. This question has also been dealt with by Clegg and Fitter (1978) who discuss the design of information system for groups.

As a measure of the effectiveness for the chosen system, it can be mentioned that throughput time in production has fallen by a factor of 3-5 depending on the product, and with maintained work-force productivity.

Conclusions

Special demands are made on production systems where materials cost accounts for a large part of the production cost in combination with rapid changes in environment. Balance and fit between the various elements are needed to obtain high flexibility at low cost, as demonstrated in the figure below.

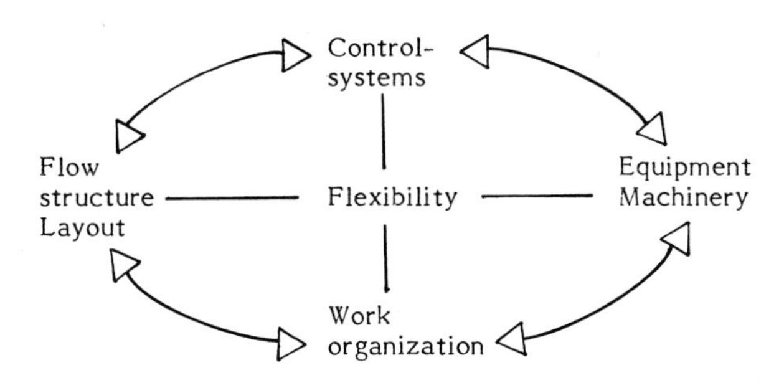

Figure 7 Interaction model

When a certain flow structure has been decided, the chosen work organization and available technical equipment must be in balance to achieve an efficient system. The balance of the control systems proved to be of central importance in this case. These systems (in this respect primarily planning and salary systems) must partly be in balance with the work organization, and partly fit well together.

List of reference

Burbidge, J.L., The Simplification of Material Flow Systems, Proceeding on the VI International Conference on Production Research, Novi Sad, 1981.

Clegg, C. and Fitter, M., Information Systems: The Achiles Heel of Job Redesign? Personnel Review, No. 3, 1978.

Eckerström, G. och Södahl, L., Ekonomisk analys av annorlunda fabriker, Working Paper 1981:11, Management Press, Stockholm, 1981.

Engström, T. and Karlsson, U., Alternative Production System to Assembly Line. A Problem Concerning Material Supply, Proceeding on the VI International Conference on Production Research, Novi Sad, 1981.

Karlsson, U., Alternativa produktionssystem till lineproduktion, Sociologiska Institutionen, Göteborgs Universitet, 1979.

Lindér, J., Inledande analys och översikt rörande elektronikmontage, Institutionen för Industriell organisation, Chalmers tekniska högskola, Göteborg, 1980.

Lindér, J., Villkor för effektiv produktionsorganisation, Institutionen för Industriell organisation, Chalmers tekniska högskola, Göteborg, 1981.

Metallindustriarbetarförbundet, Förändrad arbetsorganisation, Stockholm, 1981.

SAF, Nya arbetsformer, Stockholm, 1976.

Shingo, Study of Toyota Production System from Industrial Engineer Viewpoint, Japan Management Association, Tokyo, 1981.

THE GRAI APPROACH TO THE STRUCTURAL DESIGN OF FLEXIBLE MANUFACTURING SYSTEMS

L. PUN, G. DOUMEINGTS and A. BOURELY

Laboratoire d'Automatique GRAI, Université de Bordeaux 1, 33405 Talence Cedex, (France)

ABSTRACT

We address, in this paper, all persons interested in the structural design of FMS. However we address particularly analysts who use simulation methods. The approach suggested consists of two procedures. A progressive procedure which shows how to succeed in the design by starting with reasonable and feasible objectives, and then how to progress on the basis of experience. A decomposition procedure which comprises methods guiding the following four design steps :
a) Activity structuring and problem understanding,
b) Specification of simulation system,
c) Resolution or formal problem solving
d) Implementation of the computational process.

1. INTRODUCTION

Confidence into the merits of flexible manufacturing systems (FMS) is to-day an established matter. How to do adequateley the functional design assured at an optimal configuration of the system's configuration and of the characteristics of its constituent elements is however, still a difficult problem. Current propositions to solve the related theoretical problems are mainly of two lines :

a) Analytical methods

b) Simulation aids

1.1 Analytical methods

FMS are continuations and extensions of conventional manufacturing systems. Apparently, all the methods developped in the operational research field may be useful (Scheduling, inventory control, ressource allocation, optimal routing, queueing, statistical prediction); they are, however, insufficient in their practical application (Ref 1-4), because in FMS, the many problems are closely interrelated and these methods only yield isolated solutions to specific aspects of the system as a whole.

Several groups of researchers (Ref 5-9) have recently attempted to solve some types of scheduling problems which might occur in an FMS. They notably consider multi-product queueing inputs, and try to determine the conditions of dynamic cyclical behaviors. Interested results have been obtained, but those are too restricted to be useful for a general FMS design.

1.2 Simulation aids

Since the first works of Forrester around 1960 (System Dynamics, DYNAMO), there have been developped a great number of languages for treating "management" or "project" - type problems : GPSS, SIMSCRPT, SIMPL/1, GASP, Q-GERT and derivatives, ECSL, CAPS, SIMULA, CSS1, SLAM, etc... A short comparative bibliographical study has been made (Ref 17) in terms of learning easiness, of programming effort, portability, flexibility, processing power, compiling speed, etc... All these aids have two main components :

a) a modelling tool : linguistical or graphical, aiming to represent the activity structure of the underlying process;

b) a programming language, aiming to describe the problem at hand for computer simulation.

With respect to FMS design we notice that they offer :

1° Little facilities to represent explicitly the characteristics of the physical FMS elements : speed, power, lead capacity etc... Generally only the processing time is taken into account.

2° Little facilities to represent production scenarios, in terms of product types and production rates. Generally, the attention is concentrated on queueing problems.

3° No provision, in terms of implementing algorithmes, to sole various types of optimization problems. Even worse, the language is not adequate to write one's own optimizing algorithm.

4° Little facilities to introduce dynamic on-line decision aid procedures.

It comes out that the FMS design problem is above all to find out what are exactly the design objectives and the design framework.

2. OBJECTIVES AND FRAMEWORK OF FMS DESIGN

2.1. Present situation

Actually, most of FMS simulation studies have a limited framework and some traditional objective. The framework is formed by a set of hypothesis of the following types :

- Existing manufacturing systems, or some extensions.
- A line type FMS, ie : several machine-tools fed by a track of mobile pallets
- Random arrival of work pieces (queues)
- Some approximate knowledge of the pass-times.

The objective is to study the working feasaibility of the machines in terms of the given queues of work-pieces. There is no methodology implemented to find the optimal scheduling rules, or to determine more optimal input sequences.

Such an objective is definitely too limited in view of the ambitious aim of "high productivity" of FMS.

2.2 Comprehensive objective and complete framework

The comprehensive objective of a FMS is high productivity. To obtain it requires jointly the maximum occupation rate (100 % if possible) and a high manufacturing efficiency. The complete framework must then comprise :

a) a set of well defined products (types, variants) with their associated bills of materials;

b) a set of well defined technical processing procedures (lead times, parameter seting of machines, tools to be used etc...)

c) a set of typical production programs including for each product, finishing dates, quantities to be produced, and sequences;

d) a set of well defined characteristics of the physical elements of the FMS to be designed (number of elements, speeds, powers, loading capacities, geometric parameters);

e) well defined optimal schedulings not only for the utilization of the machine-tools, but also for that of the other physical elements (pallets, conveyors, intermediate stocks, etc);

f) well defined optimal routing programs of the work-pieces during the manufacturing processes;

g) well defined adjusting strategies, at the presence of random perturbations, either under the form of predetermined decisions, or defined in real time by automatic algorithmic computations;

h) a suitable control system (computerized) for supervising, program generating, and ajusting.

Most elements of this framework are not necessarily existing. It is therefore understandable why present studies are situated at a rather modest level. Understandable, but not excusable, if we wish to reach higher levels of FMS-designs. In fact, to reach these higher levels, we have to solve not only one FMS-design problem, but all the design problems related to the following aspects :

DP1 - Optimal sets of products including types and variants
DP2 - Optimal production programs
DP3 - Optimal technical processing procedures
DP4 - Optimal composition of the physical FMS elements (numbers, characteristics)
DP5 - Optimal scheduling and routing programs
DP6 - Optimal Computer Control System
DP7 - Optimal simulation methods and analytic procedures

It is easy to notice that many of these problems are interdependent. It seems indispensable to treat them in a comprehensive manner. The GRAI approach was developped to help with those aims in mind.

3. THE GRAI APPROACH

The GRAI approach is the utilization of (a) a set of tools, and (b) a set of methodologies, to help in the analysis and synthesis of Artificial-Intelligence Systems, particularly of those related to Industrial Production Systems.

3.1 Tools

The tools are :

a) GRAI-nets (Graphes with Results and Activities Interrelated)

b) First-order Predicate Logis

c) Category-theoretical formalism

d) HBDS : Hypergraph Based Data Structures.

The GRAI-nets are a type of graphical tools, expecially designed to structure Discrete Activities. The other three tools are associated with the GRAI-nets to increase the computability, and the information-modelling ability.

The GRAI-nets are an extension of Transfer Function, PETRI-nets, GAN, PERT and Q-GERT. The basic elements of the GRAI-nets are: (Ref. 18, 19, 20)(Fig.1).

(a)

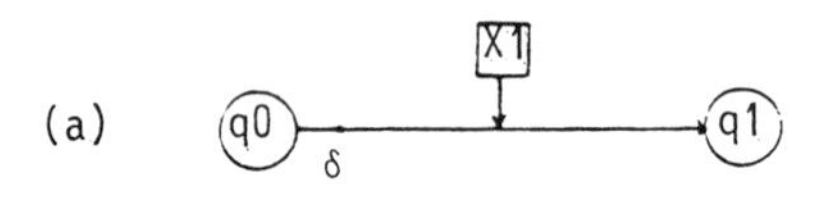

(b)

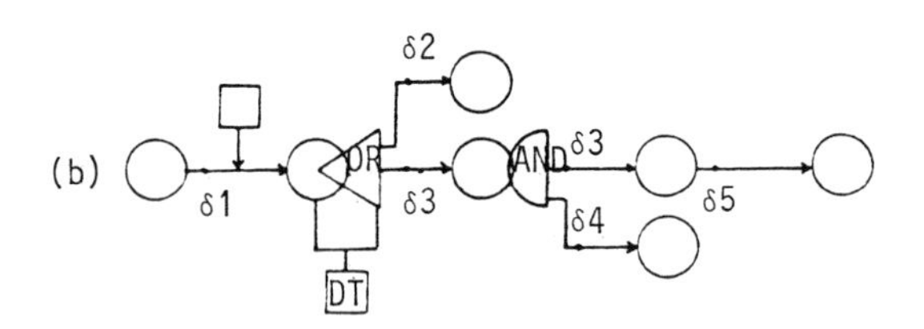

(c)

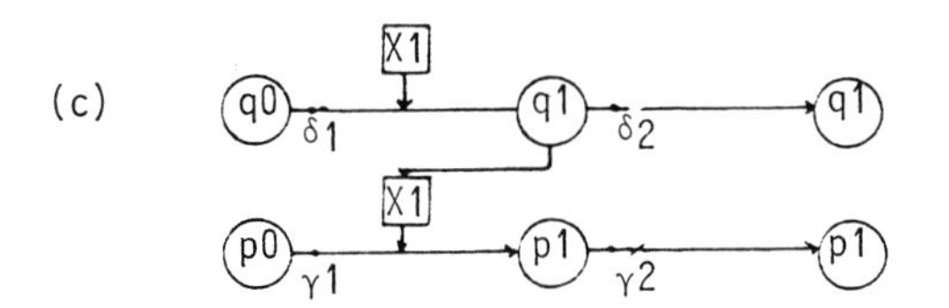

Fig. 1. Base elements of the GRAI-NETS : (a) activity model
(b) activity-chain model
(c) parallel Activity-branch model

a) Activity model : q = states, x = supports, δ : $(q0 \wedge x1) \rightarrow q1$

b) Activity-chain model : with OR and AND-successions, and DT=decision table. This model exhibits the analog and logical chain multiplicativity.

c) Parallel activity-branch model where the result q1 of the activity $\delta 1$ is used as the support y1 of the activity $\gamma 1$. This model exhibits the analog and logical branch additivity.

On the basis of these elements, one can practically model most of the complex activity structures ...

The tool "Predicate" is associated to write computer programs of the activity structures. The category-theoretical morphisms are associated to increase the solution computability. The HBDS is associated to facilitate the creation of the computer data base. All these tools have same 3 fundamental abstract concepts : states (or variable , supports (or parameters), and operation (or relation).

3.2 Methodologies

The GRAI methodologies are procedures, established on the basis of the preceding tools, and they are intended to help in the solution of the design problems of Artificial-Intelligence Systems (AIS). There are separate methodologies for each design step, and one global methodology. The following steps (derived from the conventional system-analysis procedure) are especially determined to emphasize the difficulties encountered in the design of AIS.

a) Methodology for structure understansing and problem formulation
b) Methodology for AIS specification
c) Methodology for theoretical resolutions
d) Methodology for computer implementation
e) Global progressive integration methodology for the design of complicated AIS.

4. GLOBAL PROGRESSIVE - INTEGRATION METHODOLOGY

4.1 Past Experience

We start by discussing the nature of the global methodology, which makes it easier to understand the methodologies for the individual synthesis steps. The global methodology for the development of an AIS is a procedure which takes into account its utility, its usability, its feasibility, and its developability.

An AIS (or its "fashions" name, expert system), is a computational process based on a computer language (CP/CL). The utility of a developped CP/CL can be judged in two ways : (ref. 21-24) :

- long-range utility, necessitating these types of CP/CL; covering a great number of cases, and a huge amount of intelligences,
- short-range utility, necessitating those types of CP/CL, characterized notably by their practical usability (by people with some reasonable training).

At the present time, we have two significant examples of the first type :

- APT for programming tool-path in the Numerical Control of Machine-tools,
- LISP for general problem solving.

In the case of APT, we notice that :

a) There have been many modifications for limited application areas : MINIAPT, IFAPT, EXAPT, etc ...
b) It requires many post-processors for its practical utilization (one service company has developped more than 800 post processors).
c) Introduction of NC into industry has been very slow during the last 20 years, even in industrialized countries.

In the case of LISP, we notice that :
a) It is absolutely not known in industry;
b) Even for fundamental academic researchers, it is difficult to handle.

Both APT and LISP are characterized by the largeness of their scope. Their developments are certainly very interesting for the scientific advance of mankind. We see, however, that :
a) Either it is not usable practically (LISP)
b) or it is difficult to adapt it for practical uses (APT)

The reasons are :
a) the scope of problems to be solved is not well defined,
b) the AIS is not well specified with respect to the problems,
c) the theoritical solutions are not studied according to the level of the AIS,
d) the computer implementation (language, data base, knowledge base, resolution algorithms) is not made according to the adopted theoritical solutions;
e) the educational level of practical users is not a priori considered.

4.2 The proposed methodology

The global progressive integration methodology is a procedure consisting of two subprocedures :
Procedure I - Iterative procedure
Procedure D - Decomposition procedure

The iterative procedure is the following (fig. 2) :

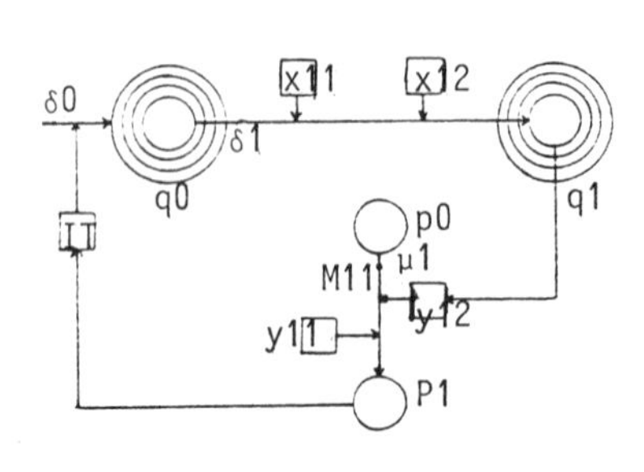

Fig. 2. Activity - elements involved in an interative procedure

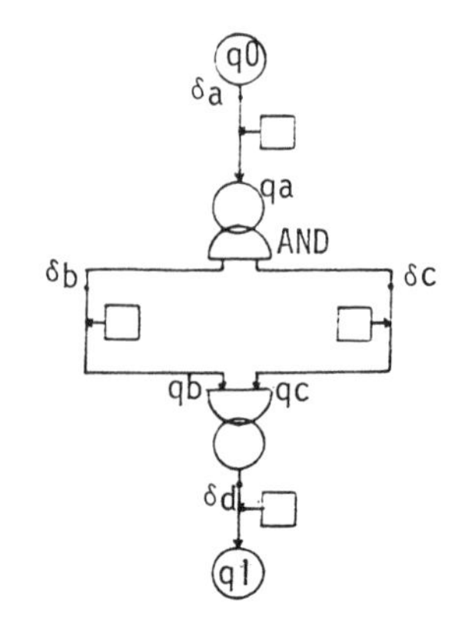

Fig. 3. Activities involved in the decomposition procedure D

I_1- Determine the minimum level utility objective Q_0, after a previous preparation δ_0 ensuing a priori the developability, the feasibility and the usability of the future AIS Q1 (Ex : a minimum size FMS, with some defined production programs).
I_2- Prepare the required knowledge, support X_{11}, within the framework of the defined objective (knowledge on technical processing, modelling techniques, resolution techniques, computer software and hardware).
I_3- Prepare the required human support (managers, workshop staffs, system analysts, computer scientists).
I_4- Develop the AIS Q1 (effectuate the operation δ_1) according to the decomposition procedure D.
I_5- Exploit the AIS (effectuate the operation μ_1 to solve current problems Po, by the practical users Y_{11}, with the aid (the support) Y_{22} coming from the AIS Q1.
I_6- On the basis of the gained experience P_1: iterate the preparation δ_0 in order to define a new utility-objective Q_0 of a higher level. Go back to I_1.

The decomposition procedure D is the following (Fig. 3).

D1 : Structure (operation δa) the activities of the FMS production system, and jointly these of the management system (MST) (if this is partially automatized FMS), and those of the AIS (see details in the next paragraphs). The aim is the understanding qa of the problems which arize during the execution of these activities.
D2 : Specify (δb) the characteristics of the future AIS according to some compromission between the desired aid (always very high) and the development possibility available (always very low). The results are the sepcification qb.
D3 : Determine (δc) the methods (analytical, heuristic) which solve theoretically all the problems. The results are the theoretical solutions qc.
D4 : Implement (δd) the specifications and the solutions into an usable CP/CL (the AIS).

5. STRUCTURING METHODOLOGY

The structuring methodology MSTR is a procedure aiming to describe and structure a set of complex activities, in order to understand clezrly the problems underlying the execution of these activities. The main confusion which arises actually in an FMS, comes from an unsufficient understanding of the relations between the objective, and the hypothesis of the framework. The MSTR will clarify this situation. The procedure is the following : Start with the initial utility objective q0 (corresponding to one of the levels of q0 defined in the global methodology I). The MSTR procedure can itself be viewed as a structure of activities (Fig. 4).

MSTR 1. Determine the required supports for the design work. Operation T1; support s1 (preferably analysts and managers). Results q1 : the names of the supports. The operation can be aided by the diagram representing the model of the design activities (Fig. 5).

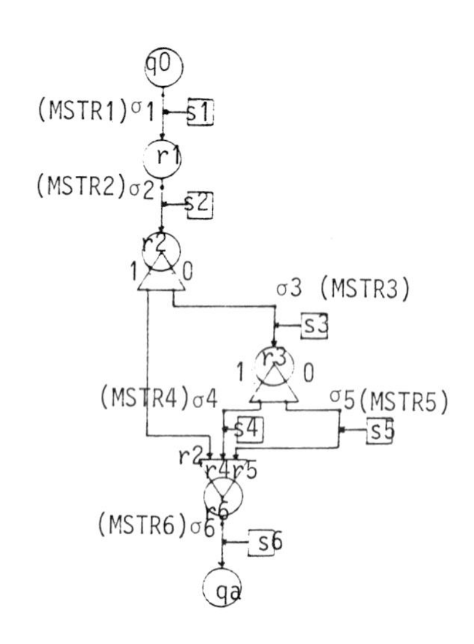

Fig. 4. Activities underlying the structuring methodology MSTR.

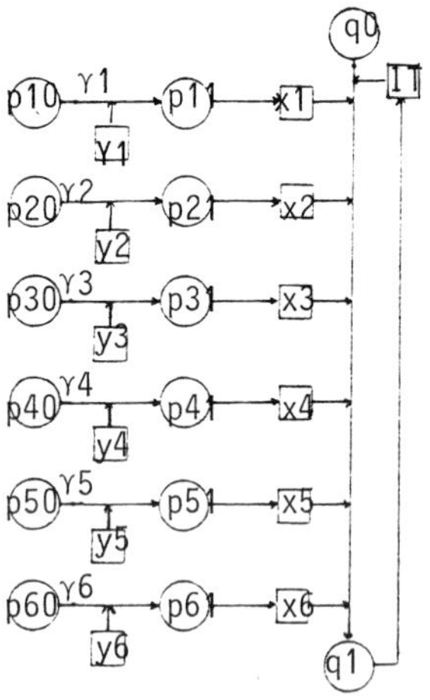

Fig. 5. Model of FMS design activity.

In the FMS-design, the supports are :

X_1 : defined products and production programs

X_2 : technical processing procedures corresponding to X_1

X_3 : characteristic of the FMS physical elements (numbers, speeds, capacities,..)

X_4 : scheduling of the operations and routing programs of the work-pieces

X_5 : computerized dynamic control system

X_6 : computerized simulation system

r_1 = name (X_1, X_2, X_3, X_4, X_5, X_6).

MSTR 2. Analysis of the availability of the supports. Operation T2; supports s2 (managers, analysts, specialists).

This analysis consists of a set of activities, the operations of which are $\gamma 1$; $\gamma 2$; $\gamma 3$; $\gamma 4$; $\gamma 5$; $\gamma 6$ (Fig. 5). The states p10, p20, p30, p40, p50, p60 are initialy states of the supports at the time when q0 is being defined. They are variables, and not necessarily equal to the set (X_1 to X_6).

The required supports y1 to y6 for the operations $\gamma 1$ to $\gamma 6$ are summarized in Table 1.

The result r2 of the operation $\delta 2$ has two values :

r2 = 0 : not available (go to MSTR3)

r2 = 1 :available (go to MSTR6)

TABLE 1 : Required supports to determine the supports of the FMS design work.

FMS - design supports	Required supports to determine the x
x1 : products and productions programs	y1 : market researches
x2 : technical processing procedures	y2 : manufacturing specialists
x3 : characteristics of the FMS physical elements	y3 : y2 + managers
x4 : scheduling and routing programs	y4 : methods + analysis, applied mathematicians
x5 : computerized dynamic control system	y5 : managers + analysts + computer specialists + users
x6 : computerized simulation system	y6 : y5

MSTR3 : Analysis of the developpability of the supports (under given conditions of quality, delay, cost and manpower).

Operation $\delta 3$, supports s2 = s3. The result r3 of $\delta 3$ has two values.

r3 = 0 : not developable (go to MSTR 5)

r3 = 1 : development of the supports.

Operation $\delta 4$, supports s4 (specialists). The results r4 of $\delta 4$ is the certainty of the development of the support.

MSTR5 : Determine a reasonable background (of framework) which is an operating hypothesis. Operation $\delta 5$, support s5 = s2. The results of $\delta 5$ is workable hypothesis r5.

MSTR6 : Determine the problems of the Design work. The imput state is :

$r6 = q_6$ OR q_d OR q_e

Operation δ, support s6 = s2. The result q_a is the set of problem statements.

In the most general case, we have : (ref. 26, 28)

- FMS operating problems :
    - state problems : program definition, schedules, routings :
- dynamic control problem : multi-synchronisation, multi coordination
- technical processing problems
- computerized-simulation problems

The aim of this methodology is to make <u>conscious</u> of the problems. The methodology has been applied during the last 4 year to analyse qualities of the Management Systems, and to specify Information Systems (More than 30 applications)..

6. SPECIFICATION METHODOLOGY (MASPEC)

6.1. Preliminary remarks

Once the problems are formulated, the next question is : what kind of AIS do we need? Assuming that the general AIS has the following constitution : (Fig.6)

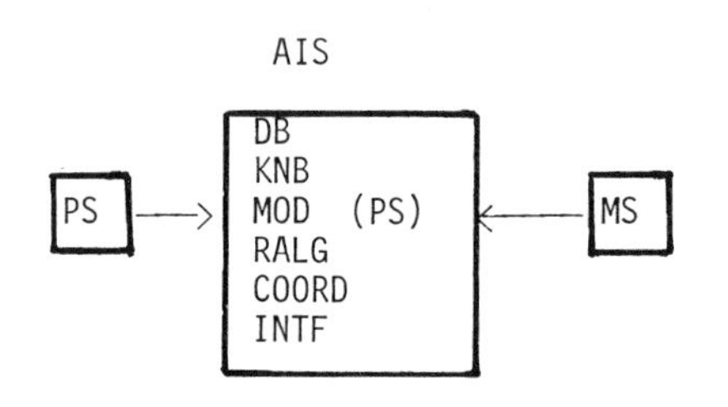

Fig. 6. General constitution of an AIS.

- Data base DB
- knowledge base KNB
- Model of the physical system MOD(PS)
- Resolution algorithms : RALG
- Coordinating procedures : COORD
- Interfaces : INTF
- Links between PS and AIS : LPI
- Links between MS and AIS : LMI
  (MS being the management od the user-system).

The specification problem is to determine the desired characteristics of these elements. Recall that the lack of adequacy of the large AIS (such as APT and LISP) is mainly due to the intricate relations between the following non-answered questions :

a) What are exactly the problems to be solved and their framework.
b) What are the intelligent operations that the MS can effectuate, and what are the aids he whish to obtain.
c) What is the constitution of the AIS to be built and how to use it.

The proposed methodology is intended to decrease (and to suppress if possible this intricary).

6.2. Proposed methodology

To facilitate the exposition, we assume known the following elements :

a) We start with a known level of utility-objective q_0 and the associated framework (production programs defined characteristics the FMS physical elements known; problem : determine their number).
b) The designs of the Computerised dynamical control system and the Computerized simulation system are similar in nature.
c) The various aid-degrees and their implications in the PS-modeling tasks and the MS on-line works are given in Table 2.

The procedure of the methodology MSPEC is the following : (Fig. 7)

Initiation : the structuration methodology MSTR yields the problem statements q_a.
MSPEC 1 : Determine, by considering Table 2, the desired aid-level. Operations: S1; results : the level t1 (eg : the evolutive level).
From t1, three activities are issued .

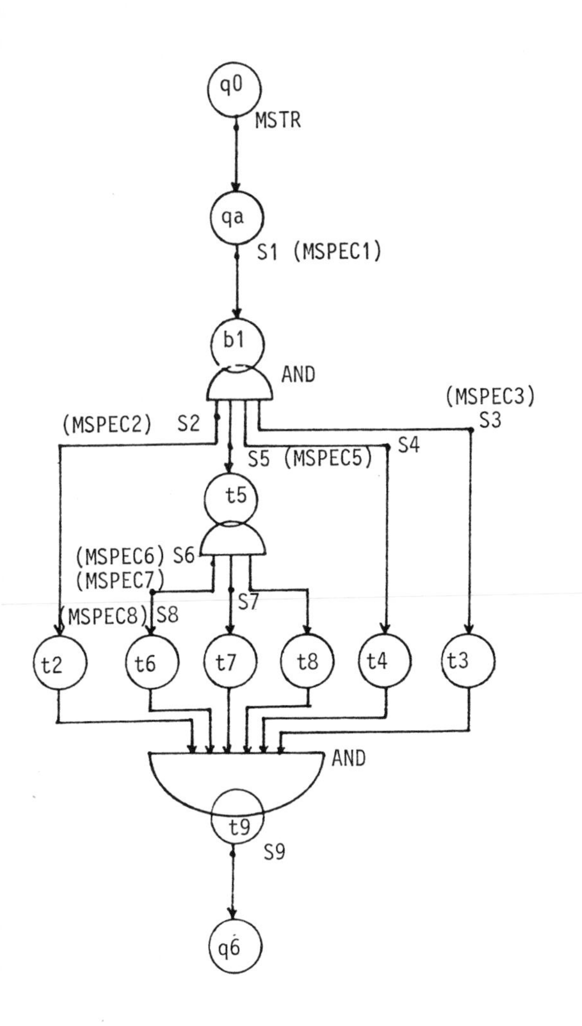

Fig. 7. Activities underslying the Specification methodology.

MSPEC2 : From the defined level of the decision-aid (t1), determine the connexions LP1 between the MAS and the AIS (eg : MS introduces the parameters and the criteria, the AIS has to compute the solution). Operation S2, result t2 = LM1.

MSPEC3 : From t1, specify the required model of the PS (eg : machine characteristics, technical processing procedure). Operation S3, result t3 = MOD (PS).

MSPEC4 : From t1, specify the required connexions between the PS and the AIS (eg : exline date collection). Operation S4 results t4 = LP1.

MSPEC5 : From t1, specify the overall background of the AIS. Operation S5, results t5 : BAKG (AIS). From t5, three activities emerge.

MSPC6 : From t5, specify the information required to constitute the data base. For this use the GRAI-tools to model the FMS-activities, the information is related to the states and the supports of these activities. Operations S6, results t6 = SPEC (DB).

MSPEC7 : From t5 specify the required knowledge to constitute the knowledge base KNB. The pieces of knowledge are the operation in the GRAI-FMS model. Operation S7, results t7 = SPEC (KNB).

MSPEC8 : From t5, specify the required algorithms to calculate the solutions (see the next section on resolution methodology). Operation S8, results t8 = SPEC (RALG).

MSPEC9: Combine the partial specification t0 from t9 with:

$t9 = t2 \wedge t3 \wedge t4 \wedge t6 \wedge t7 \wedge t8$

From t9, deduce the specification of the coordination program : t10 = SPEC(COORD) and the specification of the interface program t11 = SPEC (INTF). Combine all the specifications to form the global specification of the AIS. Operation S9, results $q6 = t9 \wedge t10 \wedge t11$.

Table 2 : Various aid-degrees and their implications

Aid degree	Implications in		
	PS modeling	Resolutions Algorithm Requirement	MS work
a) Consultation	no	no	consult and decide
b) Decision Table type	no	decision-tables	consult the decisions tables
c) Evolutive	PS model	Resolution of PS-Problems	interactive
d) Learning	PS model and Decision model	Resolution of - PS problems - decision strategy problems	Higher-level interactive
e) Complexe Automation	PS-model Decision-model Learning-model	For PS problems decision problems and learning problem	no

7. <u>RESOLUTION METHODOLOGY MRS</u> (Formal Problem Solving)

7.1. <u>Situation</u>

In FMS-design, the various problems which ask for a theoretical resolution are : (a) Generalized scheduling (in a complex machine structure); (b) Dynamic optimal multisynchronization and multi-coordination; (c) Optimal determination of the production programs; (d) Optimal determination of the technical processing (e) Optimal determination of the characteristics of the FMS-elements solving these problems imply for all of them suitable formal modelling and an iterative process. In the iterative processes, the most difficult problem to solve is to find <u>suitable inference strategies</u>.

In the continuous universes, using the analycity of the functions, gradient techniques and various associated programming techniques can be used. Un FMS-design, we are facing discrete universes, descrete both in time and in magnitude of parameters and variable : (a) Many variables are scaled in a non-ordered set (eg : tools, machines); (b) Many relations are not analytical (eg : dependancy between the quality of the machined work-pieces and the machining conditions) (even the manufacturing cycle, which is so often used as a FMS design criterium); (c) Some local inference tactics are known, but in an heuristic and non-correlated manner (eg : if we are too close to the due-date, we put on twice the man-power). In terms of the AIS (namely Expert Systems), we have a large combinaterial problem where pieces of inference knowledge are presented piecemal fashion (Ref. 21-24). Still, we are rather poor in dealing with these inference knowledge in the case of FMS-design.

7.2. The proposed methodology

Basic philosophy : Under the framework of a given utility objective q0 (Fig.3) and a clear formulation of the problems qa, reduce the dimensionality of the variables (states, supports) simplify their scaling, construct and identify the operators, transform the multi-criteria and the multi-contraints into sequential forms so that: (Ref. 25,27) (Fig. 8).

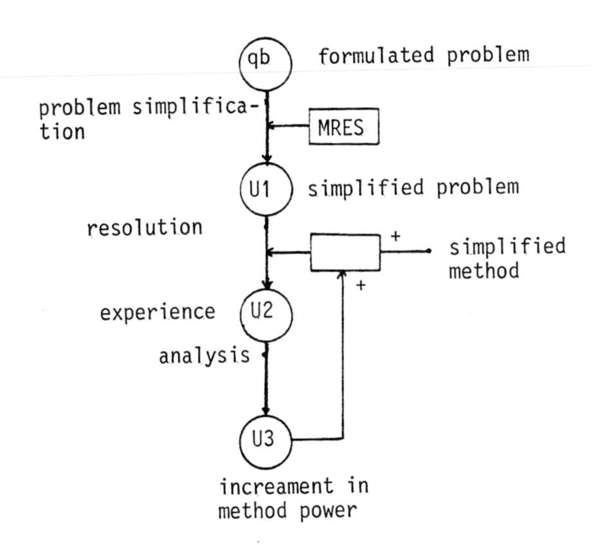

Fig. 8. Progressive resolution methodology

(a) The problem so stated has a meanning with respect to the adopted objective-level,
(b) Inference tactics can be established and the problem solved with a reasonable amount of effort.
(c) The resolution experience can be used to construc a feasible resolution space for more complex problem-formulations. To facilitate exposition of the methodology, we use the example of a "Multi-synchronizing control structure" (Fig. 9) where we have : Branch B = manufacturing activities, Branch A : procurement activities : (YjB, L, E) = decisional activities. The methodology is the following.

(Finality-oriented procedure).

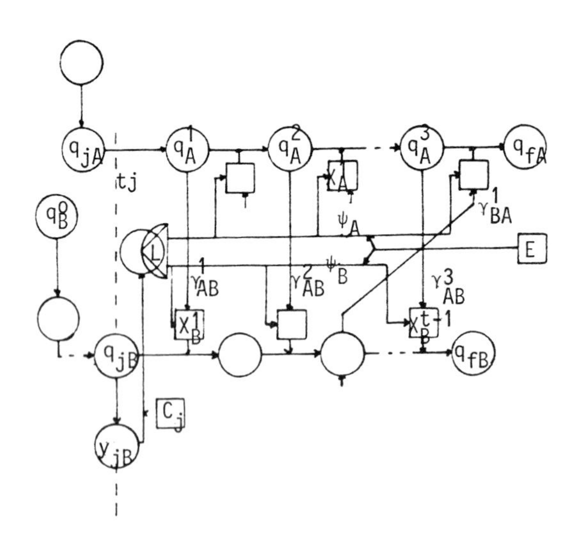

Fig. 9. Multisynchronizing Control

MRES1: Locate the main finality, use a minimum of components (QfB = final products adopt qyB = delai).

MRES2: Use the simplist scaling having (qfB ε (U,1), 1 if qfB < qr prescribed quantity, otherwise 0).

MRES3: Determine the most important variables (and their simplist scaling) which influence the value of QfB (q_bf-1) = right quantity or not, X_Bf-1 = machine present or not).

MRES4: Identify the corresponding operator γ^{f-1}(in the form of a morphism).

MRES5: Proceed in similar way through Branch B, backwards until to reach the initial state qB.

MRES6: Determine the result-support interrelations between Branch B and Branch A (the pairs q_A^1 - X_B^1, q_A^2 - X_B^2, etc).

MRES7: Identify the corresponding morphism : (γAB^1, γAB^2, etc.)

MRES8: From the related states in the Branch A, proceed to identify the B-operators in a similar way.

MRES9: From branch to branch, define suitable variables, identify all the operators of the other branches of the considered structure.

MRES10: Locate significant states where perturbations might happen (eg : qjB).

MRES11: Determine the simplist adjusting criteria (Cj = cost).

MRES12: Determine the influencing variables (qA, qB, XA, XB).

MRES 13: Identify the morphisms of the relations Cj (XA, XB, qA, qB).
MRES 14: Establish the tactics Ø according to the situation YjB, to the criteria Cj, to yield the adjusting policy L. This is the most difficult part. All the preceding simplifications aim at randering this operation easier. This methodology is easily generalized to the other types of problems mentionned in paragraph 7.1.

8. IMPLEMENTATION METHODOLOGY - MIMPL

8.1. Situation

The implementation problem, for the special objectives of FMS-designs, is to create a special computer language FMS-CL, and a special computational process FMS-CP, satisfying the following criteria :

(a) Computer-criterium : the computer model for FMS-designs must be portable, ie: implementable not in one, but in many computers.

(b) User-criterium : the computational process must be usable by commun users within the adopted level of the FMS-design objective.

A preliminary confusion must be clarified. One often wonders who must do this work? Computing professionals, or Engineers having a restricted trainning in computing sciences ? This is a bad question. The correct one is : What must be done? Instead of who must do it ?

8.2. Proposed methodology

Basic philosophy : (a) Transform the various internal models of the various types of informations, operations, and theoretical resolution methods into suitable "External models". This is done by the GRAI-net tool, and the associated tools. (b) Create a special computer language based on a class of abstract concepts common to the External models and to user understanding. This is done by the GRAI-FMS-CL (Ref. 16).

The methodology is the following (Fig. 10).

MIMPL1: Coherence analysis. Summarize the problem formulations qa resulting from MSTR the AIS-specifications q6 resulting from MSPEC, the resolution methods qe resulting from MRES. Analyze their coherence with respect to the adopted objective level q0. If coherent, split qa into :

qa' = (information classes on states, supports)

qa" = (physical decisionnal operations)

qa"'= (information on activity-structures).

Split qb into :

qb' = the part on DB, KNB, RALG, MOD(PS).

MIMPL2: Information modeling. Transform (operation M1) information q'a into an External model qm1, use the tool y1 = HBDS (Ref.30,31).

MIMPL3: Operation modelling. Transform (operation M2) operations qa" into an External model qm2, use the tool y2 = predicates (Ref. 32).

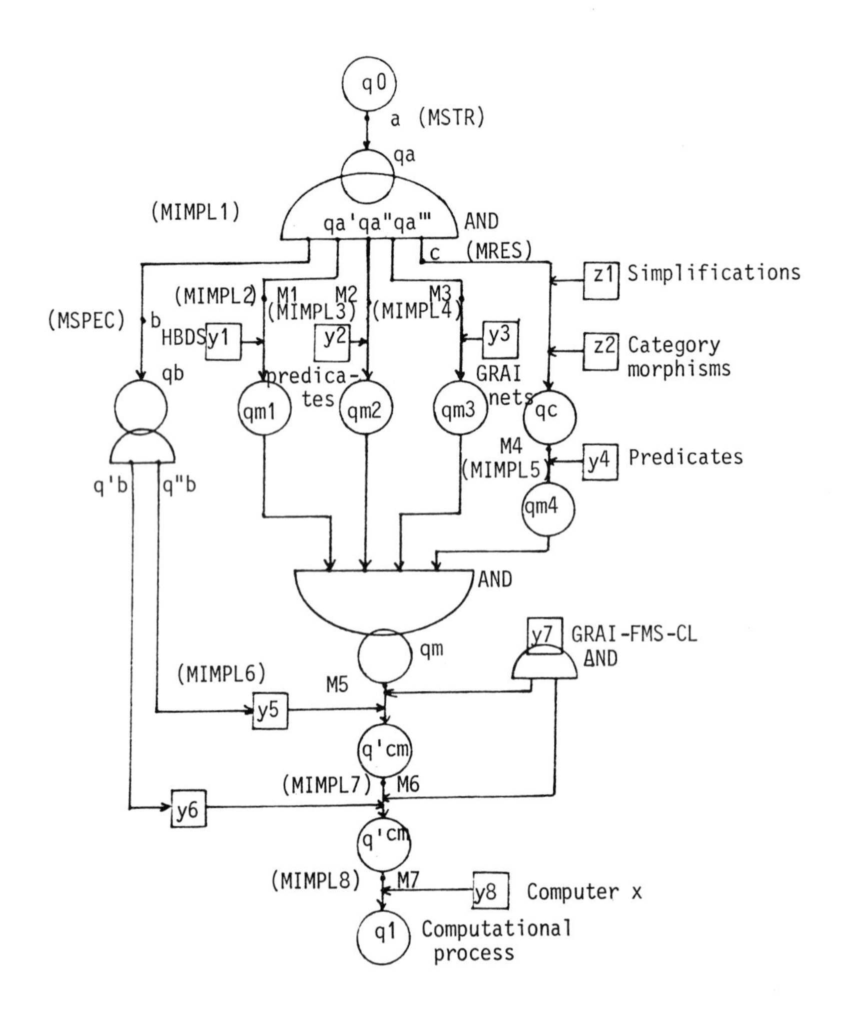

Fig. 10. Activities underlying the implantation methodology

MIMPL4: Structure modelling - Transform the activity structure qa''' (operation M3) into an External model qm3, use the y3 = GRAI-tools.

MIMPL5: Algorithm modelling - Transform the methods qc into an external model qm4, use the tool y4 = predicates.

MIMPL6: Part programming. Transform (operation M5 the external models qm1 to qm4 into computer models, use the tool y7 = GRAI-FMS-CL. Use the support: Y5 = q"6. The résult q'cm comprises DB from qm1, KNNB from qm2, MOD (PS) from qm3, and RALG from qm4.

MIMPL7: General programming. Complete the part-computer model q'cm with the creation of COORD and INTF. Operation M6. Use the support Y7. The result is the complete computer-model qcm.

MIMPL8: Final conversion. For a given computer X, convert (operation M7) qcm into the Computational process q1 = CP, with a language exploitable on the computer X = Y8.

9. CONCLUSION

In this paper, we have presented the GRAI-approach to the design of flexible manufacturing Systems (FMS). An FMS is taken in its complete sense, including the physical elements, the computer control system, and the computer simulation system. The approach is a procedure consisting of two sub-procedures : (a) the iterative methodology, starting the design on a feasible simple level, and progressively integrate the experience to advance to higher levels; (b) the decomposition methodology consisting of four steps : (b1) structure the activities to yiels a clear problem understanding; (b2) specify the AIS desired characteristics by deriving a suitable aid-degree from a feasible background; (b3) determine the resolution methods by adopting compatible simplifications, and (b4) implement the AIS computational process by establishing suitable external models of information, operations, structures and algorithms, and by use of a specially developped computing language GRAI-FMS-CL. Most of the mentionned tools are existing or have been created. Some partial experience has been oblained, and improvements are on the way.

REFERENCES

1 L. Pun, Couplage entre les problèmes et les méthodes de gestion automatisée de production, Journées de l'AFCET, 4 juin 1974, Paris.

2 L. Pun, Theorics and facts in production management problems, Invited paper, Internet IV, August 1976, Birmingham.

3 L. Pun, Approche méthodologique de modélisation en vue de la maîtrise assistée de la production, Congrès AFCET, nov. 1977, Versailles.

4 L. Pun, G. Doumeingts, J. Grislain, D. Breuil, A survey on Computer-aided planning in complicated production systems, Third IFAC Symposium on System Approach for Development, Nov. 24-27, 1980, Rabat, Morocco.

5 G. Bel, J-B. Cavaille, D. Dubois, Outils de conception et ateliers flexibles d'usinage,Revue Générale d'Electricité, Nov. 1982, 742-746.

6 S.B. Gershwin, J.G. Kimenia, Multi commodity network flow optimization in flexible manufacturing systems, Report ESL FR-834-2, Electronic Systems Laboratory, MII, Cambridge, Mass., April 1980.

7 J.G. Kimenia, S.B. Gershwin, Analgorithm for the computer control of production in a FMS, 20 th IEEE Conf. on Decision and Control, San Diego, Dec. 1981.

8 J. Erschler, F. Roubellat, V. Thomas, Une approche pour l'ordonnancement en temps réel d'ateliers, Congrès AFCET Automatique, Nantes, Octobre 1981.

9 J. Erschler, Lancement de produits dans un atelier flexible, Journées ARA, 28-30 Sept. 1982, Poitiers.

10 J.A.A.B. Pritsker, The GASP IV simulation language, John Wiley, 1974.

11 A.A.B. Pritsker, Modeling and Analysis using Q-GERT network. Halsted Prem 1977.

12 A.A.B. Pritsker, PREDEN CD, Introduction to Simulation Language, Users Manual, University of Birmingham, June 1980.

14 A.I. Clementson, ECSL and CAPS, Extended Control and Simulation Language and Computer Aided Programming System, Detailed Reference Manual, May 1980, Uni-

versity of Birmingham.
15 Ch. Bérard, A. Bourely, D. Breuil, Simulation d'ateliers, ASME 82, Juillet 1982, Vallée de Chevreuse.
16 Ch. Bérard, Contribution à la conception de structures logicielles pour le pilotage d'ateliers, Thèse Dr-Ing. Jan. 1983, Université de Bordeaux 1.
17 A. Bourely, Méthodes d'évaluation des performances pour la conception d'ateliers flexibles, Etude Bibliographique, Laboratoire d'Automatique GRAI, Université de Bordeaux 1, Sept. 1982.
18 L. Pun, Computer assisted static and synamic planning for productions activities, Automatica, March 1981.
19 L. Pun, G. Doumeingts, J. Grislain, D. Breuil, Modelling organized discrete activity control, IFAC Congress VIII, Kyoto, August 24-28, 1981.
20 L. Pun, Réseaux GRAI et application à la conduite automatisée de la production, R.G.E., n° 11, Nov. 1982.
21 National Aeronautics and Space Administration Headquaters, An overview of Expert Systems, NBSIR 82-2505, May 1982, Washington D.C. 20546.
22 J.L. Laurière, Représentation et utilisation des connaissances, Techniques et Sciences Informatiques, n° 1, Janv. 1982, pp 25-42, n° 2, Mars 1982, pp 109-133.
23 N.J. Nilsson, Principles of Artifical Intelligence, Springer Verlag, 1982.
24 L. Pun, Artificial Intelligence Systems and Applications to Production Systems, Ed. Tests, 1984.
25 J. Grislain, L. Pun, Ch. Bérard, Information aggregation to the degree of decision aid in production management, IFIP APMS "82", August 24-26, 1982 Bordeaux
26 L. Pun, G. Doumeingts, Predictive complicate coordinating planning, Pacific Operation Research Symposium, Nov. 16-19, 1982, Singapoore.
27 L. Pun, Duality analysis in Computer-decision process, IFAC Workshop on CAD/PD, June 22-24, 1982, Ankara.
28 L. Pun, G. Doumeingts, GRAI-nets and their applications to CAP problems, IFAC, Workshop on CAD/PD, June 22-24, 1982, Ankara.
29 F. Bouille, Un modèle universel de banque de données, simultanément partageable, portable et répartie, Thès d'Etat, Université de Paris VI, avril 1977.
30 F. Bouille, The Hypergraph-Based-Data-Structure, and its applications to data structuring and complex-system modeling, Lecture notes, University Paris VI, October 1979.
31 L. Pun, D. Breuil, Calculabilité dans la conduite des activités discrètes, IFAC Congress IX, July 2-6, 1984, Budapest.

HEURISTICS FOR LOADING FLEXIBLE MANUFACTURING SYSTEMS

KATHRYN E. STECKE and F. BRIAN TALBOT

Graduate School of Business Administration, The University of Michigan,
Ann Arbor, Michigan (USA)

ABSTRACT

The flexible manufacturing system (FMS) is an alternative to conventional discrete manufacturing processes that permits highly automated, efficient, and simultaneous machining of a variety of part types. In managing these systems, technological requirements indicate that several decisions must be made prior to system start-up. To this end, previous research has defined a set of FMS production planning problems. The final production planning problem is called the loading problem, which is described as follows. Given a set of part types chosen for immediate simultaneous production, allocate the operations and associated tooling of these part types among the machines subject to the capacity and technological constraints, and according to some loading objective. This problem has previously been formulated as a nonlinear mixed integer program for several loading objectives. Although it has been shown that the nonlinear MIPs are solvable on large computer systems, real-time FMS control requirements and the typical availability of minicomputers in shop environments make it impractical and cost inefficient to optimally solve the loading problem in many plants today.

As a result, the authors develop several heuristic algorithms that provide good solutions to various versions of the FMS loading problem. We expect that these rules can be executed in essentially real-time on minicomputers available today.

INTRODUCTION

The development of highly automated flexible manufacturing systems (FMSs) has created new opportunities for the efficient manufacture of component parts in the metal-cutting industry. The effective use of these systems, however, requires the solution of new and complicated production planning and control problems. In an effort to make these problems tractable, Stecke [1983] has devised a hierarchical scheme comprising five production planning problems which must be solved prior to system operation. A brief description of these appears in Table 1. The primary purpose of this paper is to present heuristic solution procedures for one of these problems, the FMS loading problem.

The loading problem is one of deciding how individual machines are to be tooled to collectively accomplish all manufacturing operations for each part type that will be machined concurrently. A solution to this problem specifies the cutting tools which must be loaded in each machine's tool magazine before production begins, and hence, the machine or machines to which a part can be routed for each of its operations. Since a variety of products (parts) can be manufactured simultaneously on an FMS, where each part has its own, potentially unique, set of required operations, loading becomes a combinatorial problem. Some of the characteristics which make this a difficult combinatorial problem to solve include the possibilities that:

i) some particular part operations may be performed on any of several different types of machines;

ii) operations could then require different processing times on various machine tools;

iii) different part operations may be able to use some of the same cutting tools; and

v) tools, measured individually and collectively, require various amounts of space (slots) in fixed-size, limited-capacity tool magazines.

TABLE 1

Production Planning Problems.

1. Part Type Selection:
   From a set of part types that have production requirements, determine a subset for immediate and simultaneous processing.

2. Machine Grouping:
   Partition the machines into machine groups in such a way that each machine in a particular group is able to perform the same set of operations.

3. Production Ratio:
   Determine the relative ratios at which the part types selected in problem (1) will be produced.

4. Resource Allocation:
   Allocate the limited number of pallets and fixtures of each fixture type among the selected part types.

5. Loading:
   Allocate the operations and required cutting tools of the selected part types among the machine groups subject to technological and capacity constraints of the FMS.

Although it is possible to model these characteristics as nonlinear mixed integer programs (Stecke [1983]), it can be time and cost prohibitive to optimally solve the resultant loading problems that are large, despite the existence of an efficient branch and bound procedure that can solve the non-linear integer problems associated with one of the 5 FMS loading objectives of interest here (Berrada and Stecke [1983]). Hence, there will be a need for fast heuristic procedures that give good, if not optimal, solutions for large scale FMS loading problems.

The loading problems addressed in the first section assume that the grouping problem ((2) of Table 1) has been solved. Stecke [1981] introduced the notion of grouping machines as one way of simplifying overall production planning and control of FMSs. Grouping also automatically provides machine redundancy, which is very useful during machine breakdown situations. A machine group is composed of machines that are tooled identically so that they can individually perform the same operations. Typically, machines in a group are of the same type and are identically tooled. Through the use of closed queueing networks to model FMSs, Stecke and Solberg [1982] found optimal grouping patterns and associated optimal allocation ratios, which indicate the amount of work (operation processing time) which should ideally be assigned to each machine group, to provide maximum expected production. Knowing how machines should be grouped affects other aspects of production planning, but specifically, simplifies the loading problem both by reducing the tooling options and by reducing the size of the problem to be solved. In ∞2, FMS loading heuristics are suggested that also group the machine tools.

TABLE 2

Alternative Loading Objectives.

1. Minimize part movement between machines, or equivalently, maximize the number of consecutive operations for a part to be processed by the same machine;
2. Balance the workload (total processing time) per machine on all machines;
3. Balance the workload per machine for a system configured of groups of machines of equal sizes;
4. Unbalance the workload per machine for a system of groups composed of unequal numbers of machines;
5. Duplicate certain operation assignments.

Several loading objectives are considered in this paper and are indicated in Table 2. Each might be applicable in different manufacturing situations. In some systems, several objectives may apply.

1. LOADING HEURISTICS WITH GROUPING

In this section, several loading algorithms are described for the situations where the grouping problem has already been solved. That is, we know how many groups there are, the sizes of each machine group, and which machines are in each group. The solution to the loading problem will define precisely which operations, and hence tooling, will be assigned to each machine group. There are several machines of each type. (If there is only one machine of each type, then the loading problem that is solved in this section becomes trivial). We initially assume that each operation can be accomplished by only one machine type. This assumption can easily be relaxed.

1.1 FMS Loading Algorithms for Minimizing Part Movement

The first two algorithms are designed to minimize part movement through an FMS. This objective is especially important in a system having relatively high travel or pallet positioning times (see Stecke and Solberg [1981]) or if the material handling system is a bottleneck. The first algorithm approaches the problem by examining consecutive operations sequentially, whereas the second, pre-groups consecutive operations before loading. Throughout, the algorithms are presented in increasing order of complexity.

ALGORITHM 1

1. Taking each part type in numerical order, assign each operation consecutively to the lowest numbered machine tool of the correct type which has magazine capacity available for the tools required for the operation.
2. Continue assigning operations until all have been allocated.

This is a simple application of the first-fit bin packing heuristic (see Johnson [1973]) and involves very little computational effort beyond feasibility testing. At each tool magazine capacity test, common cutting tool and tool slot overlap checks as well as corresponding adjustments, should be made to determine feasibility. A potential drawback of this simple, naive approach is that the resulting solution will likely not conform to given commonly-tooled machine grouping goals, related to total assigned processing times.

ALGORITHM 2

1. For each part type, group maximally into "operation sets", consecutive operations which require the same machine type. Calculate the number of magazine slots required for each operation set.

2. Calculate a priority index for each operation set, and assign operation sets to machines of the correct type according to this index. Several possible prioritizing schemes are:
   (a) assign operation sets to the lowest numbered machine possible according to the index: "largest number of tool slots required" first. This is a variation of the first-fit-decreasing bin packing heuristic.
   (b) assign operation sets according to the index "largest number of tool slots required" first, but to the machine having the cutting tools already in its magazine which will most reduce the number of slots actually needed by the operation set being assigned.
   (c) assign operation sets to the lowest numbered machine possible according to the index: "largest number of operations in a set" first.
   (d) assign operation sets according to the largest value of the ratio: (number of operations in a set)/(number of additional tool slots required). This rule is designed to assign as many operations as possible at the lowest cost in terms of additional tool magazine slots needed.

These heuristics of Algorithm 2 will, like Algorithm 1, probably give solutions which do not conform to ideal groupings of machines as provided by closed queueing network analysis. In addition, if the use of maximally grouped sets of operations does not lead to a feasible solution, then alternative methods must be devised to define operation sets. This could be a difficult problem, although as Stecke [1981] has suggested, a starting point for defining sets can be provided by the L.P. relaxed solution to the nonlinear I.P. statement of the FMS loading problem. The L.P. solution could be adjusted heuristically, to conform to the integrality requirement while remaining feasible.

1.2 Loading Procedures for Balancing and Unbalancing Objectives

Closed queueing network analysis provides idealized groupings of identically tooled machines as well as corresponding optimal group workload allocation ratios (Stecke and Solberg [1982]). In general, the analysis proves that for balanced configurations of grouped machines, expected production is maximized by balancing the assigned workload per machine. Alternatively, better machine configurations are unbalanced, and in these situations, expected production is maximized by a specific unbalanced allocation of work.

The following heuristics are designed to assign operations to machines in an effort to meet these optimal allocation ratios. These ratios were developed to provide guidelines on how to allocate work to machines to maximize expected system productivity as measured by the amount of processing time which can be completed in a given period of time.

Before the following loading algorithms are implemented, it is useful to obtain an estimate of the maximum workload per machine. This is accomplished with the following calculations:

i) Sum the operation processing times, weighted by the part production ratios, of all operations for all part types (that will be simultaneously produced by the FMS over the next time period) that require a particular type of machine. Do this for each type of machine. (Part production ratios can be either provided by a production plan, or can be analytically derived ratios designed to maximize system productivity.)

ii) Divide each of these sums by the corresponding number of machines of each type to obtain an estimate of the workload per machine, if the workload were balanced.

ALGORITHM 3

1. Order machine groups within each machine type by nonincreasing number of machines in each group (as previously solved by the grouping problem).

2. Order operations, for all part types that shall be simultaneously machined, which require the same machine type, by nonincreasing processing time.

3. Assign the first operation from each of the lists developed in Step 2, to the first machine group of the correct type.

4. The assignments of the remaining operations depend on whether the machine tools of a given type are organized into groups of equal or unequal size:

   (a) Equal Size Machine Groups (requires the Balancing Loading Objective).

   It is necessary to assign operations from their ordered lists to their corresponding required machine types. Thus, the following allocation procedure is repeated for each list. Suppose there are L machine groups for some list. Consecutively assign the first L operations from this list to the L machine groups. Consecutively assign the next L operations, but in reverse machine group order. For example, machine group one will contain operations 1, 2L, 2L + 1, 4L, 4L + 1, etcetera. Machine group two will contain operations 2, 2L - 1, 2L + 1, etcetera.

(b) Unequal Size Machine Groups (requires the Unbalancing Loading Objective)

The following procedure is repeated for each operation list as they were defined in step 3: Corresponding to each list is a set of machine groups which were ordered in step 2. Suppose these groups are labeled: $\ell=1,\ldots,L$, with the number of machines in the ℓ-th group denoted by M_ℓ. The allocation rule is to assign the next M_ℓ operations to the ℓ-th group, for $\ell=1,\ldots,L$. When $\sum_\ell M_\ell$ operations have been assigned, continue the process with $\ell=1,2,\ldots,L$. Repeat until all operations have been assigned.

Figure 1 further illustrates the application of Algorithm 3(a). Here, nine machines of the same type have been placed into three equal sized groups. Operations one to nine have been allocated thus far, as indicated by the numbers in the rectangles. The height of each rectangle is proportional to the operation processing time. The rationale for reversing the allocation across groups, in an effort to balance workload, is evident from Figure 1. If the processing times are highly variable, then it may be worthwhile to deviate from the rigid allocation procedure by skipping a group that appears to have an excess workload, or allocating an extra operation to an underloaded group, or by making adjustments (i.e, pairwise exchanges) after an initial solution has been found.

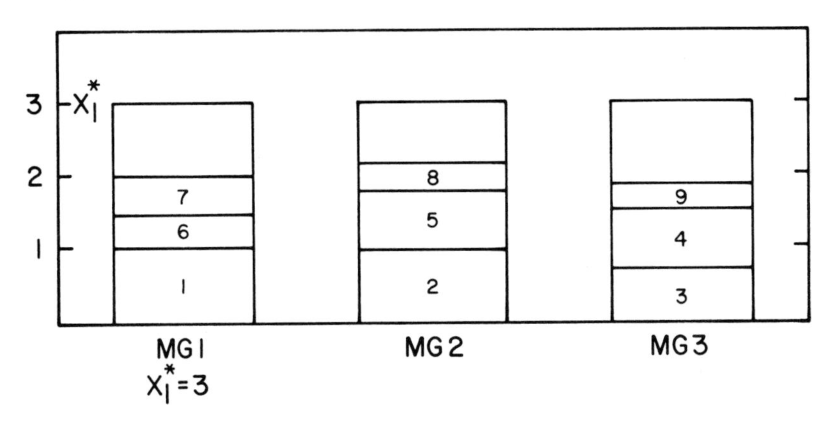

Figure 1. Three Equal-Sized Groups Containing Nine Machines.

Figure 2 illustrates Algorithm 3(b). There are seven machines of the same type placed into three groups of four, two, and one machine in each group. Operations one to 14 have been assigned thus far, as indicated by the numbered rectangles. The height of each rectangle is a measure of the operation processing time. The X_ℓ^* values are hypothetical optimal allocation ratios which would usually be obtained from using the closed queueing network model. These ratios are guidelines which could be used to measure the "goodness" of a particular heuristic (as well as an optimal) solution. Consistent with the relative magnitude of these ratios, the heuristic procedure described aims to assign slightly more than an average amount of processing time to the larger groups and slightly less than average to the smaller groups.

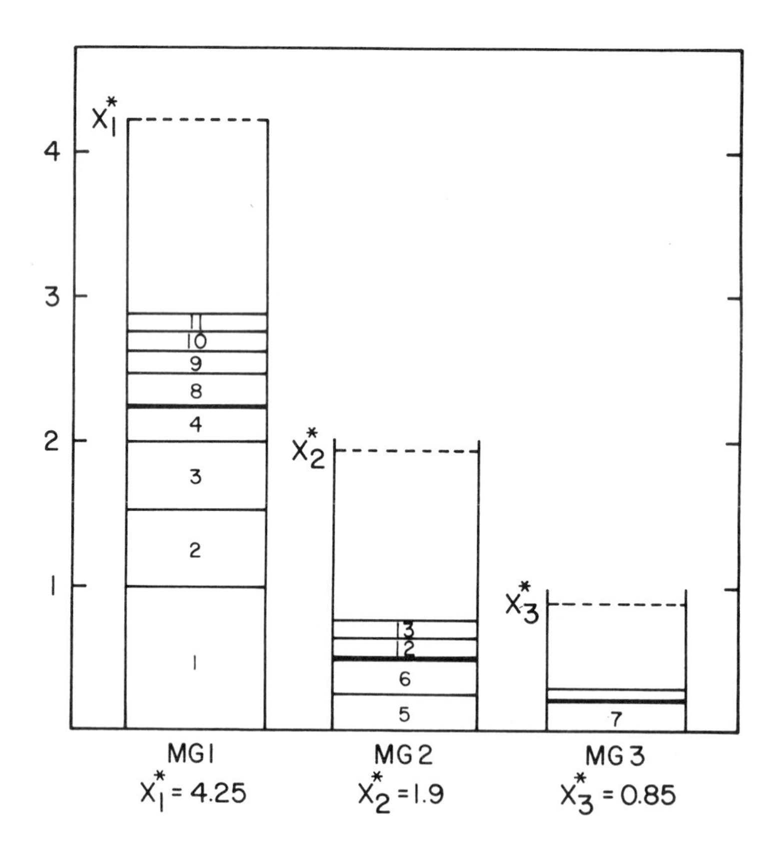

Figure 2. Seven Machines Partitioned Into Unequal Sized Groups.

A more comprehensive example demonstrating Algorithm 3 will now be given. Figure 3 illustrates a 13-machine manufacturing system containing three types of machines which have been arranged into six groups. Optimal allocation ratios X_ℓ^* are specified for each group. For each type of machine, the groups have been ordered such that the largest groups are first, according to Step 1.

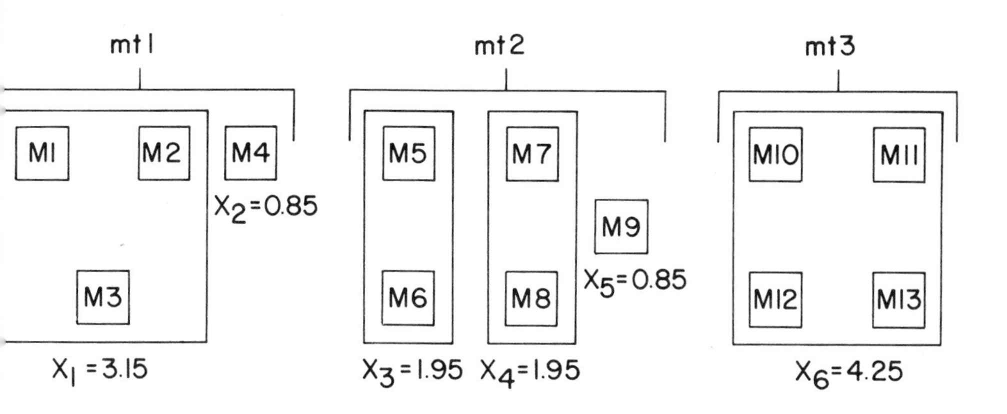

Figure 3. Thirteen Machines of Three Machine Types.

Table 3 contains the ordered operations for all parts and the type of machine required. As specified by Step 2, operations have been ordered by largest processing time first.

TABLE 3

Operations Ordered According to LPT First.

Operations	1	2	3	4	5	6	7	8	9	10	11	12	13	...
Machine Type	1	2	2	1	3	1	2	1	2	1	2	3	3	...

Since the groups for machine type 1 are of unequal size, Algorithm 3(b) is followed. The number of operations assigned to each group is equal to the number of machines in each group. Then the process is repeated until all operations that require machine type 1 have been assigned. See Figure 4.

There are three groups of type 2 machines. Since groups three and four have equal numbers of machines and group five has a different number, a combination of Algorithm 3, Parts (a) and (b), is applied. Groups three and four will be assigned operations on the forward and reverse pass. Group five will be assigned an operation at the end of the reverse pass, as displayed in Figure 4. Machine type three has only one group of four machines, so all operations are simply assigned to it in order.

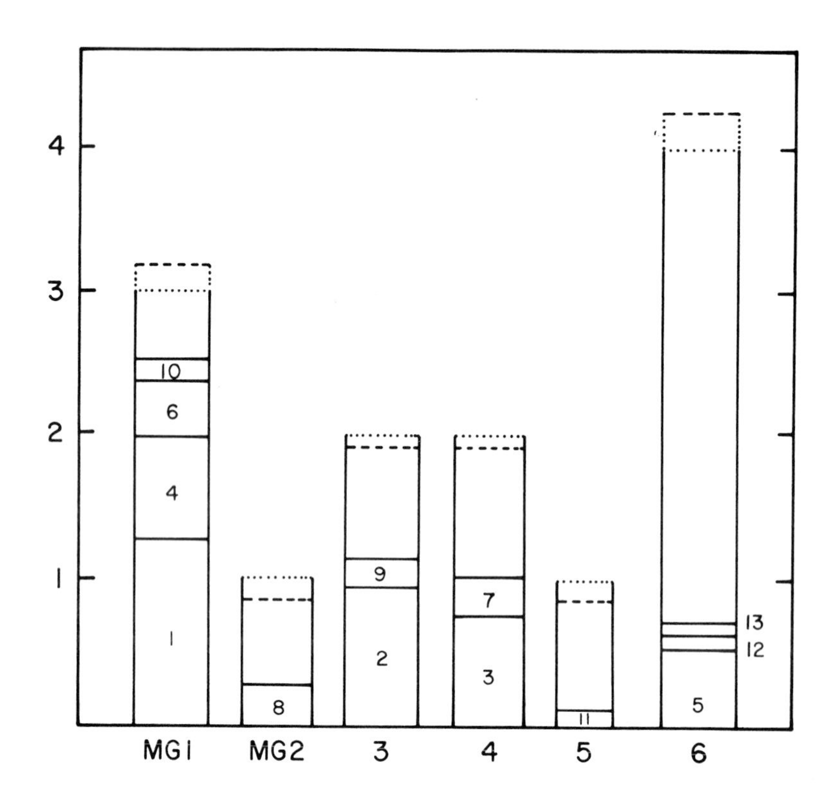

Figure 4. Allocation of Operations for the Thirteen Machine FMS.

2. LOADING HEURISTICS WITHOUT PRIOR MACHINE GROUPING

In this section, no knowledge is assumed about optimum machine grouping or allocation ratios. Machine grouping, if any, will result from the loading process.

It is assumed that there exists a set of part types ready for production on the FMS. Each part requires a predetermined sequence of machining operations, each of which must be performed by a specified machine type and machine tool. A batch of one or more identical parts that are to be made in the same production period will be referred to as a job. Jobs are assigned a priority index consistent with the management objectives of the FMS. For example, the index may be based on job due date, processing time, revenue, cost, importance of the job's customer, and so on. Loading analysis is accomplished near the end of every production period, and will specify how tools are to be allocated among the individual machines for the ensuing production period. The end of a production period does not have to correspond to the completion of all jobs in the FMS. The following algorithm uses information about the current state of the system, whether the system is empty or has jobs in process. Hence, this loading procedure could be applied to a FMS operated like a job shop.

ALGORITHM 4

1. Rank jobs in decreasing order of importance, where importance is specified by job priority indices. Call this ranked list the job arrival list.

2. For the most important job on the arrival list, sum the part processing times for operations requiring like machine types. Do this separately for each machine type. Multiply each of these sums by the number of parts in the job. These products indicate the total processing times required of each machine type to complete the entire job.

3. Check machine time feasibility. For each machine type, compare the time required by the job and the time remaining on the machine type. A job is said to be machine time feasible if the time required by the job is less than or equal to the time available for all machine types. Otherwise, the job is said to be machine time infeasible.

4. Check tool feasibility. For each machine type, compare the tool requirements and tool availability in terms of tools already on the machine, and unused tool magazine capacity. If the tools are already in place or if sufficient magazine capacity exists, then the job is tool feasible.

5a. If the job is simultaneously machine time and tool feasible for each machine type, it can be loaded into the system. This means removing the job from the available list, and updating the machine time and magazine availabilities to reflect the additional work this job places on the system.

5b. If the job is either machine time or tool infeasible, the job is considered for splitting, either by splitting the batch size, or by splitting the operation sequence. Either type of splitting will likely result in inprocess inventory, and a loss of batch contiguity. Reducing the batch size, of course, has no effect on tool infeasibility, whereas splitting operations can reduce both machine time and tool infeasibility. The desirability of splitting would involve these trade-offs as well as the potential attendent loss of machine utilization if no splitting were permitted.

6. Revise priority indices for jobs remaining on the arrival list to reflect the current status of the system, and to include any new job which may have been added to the list. Repeat steps 1-6 until no jobs remain on the waiting list, or no additional jobs, or portions of jobs (from splitting), can be loaded.

In step 5, partial machine grouping may be accomplished or avoided depending upon what is desired. Grouping requires duplicate tooling, and hence, may be uneconomical. However, grouping reduces the disruption machine or tool breakdowns may have on the completion of a particular job, it effectively increases the production rate of the job, and it may be required to avoid job splitting. To illustrate how partial grouping can be accomplished, consider a job with a batch size of N parts. Suppose that each part requires four operations on machine type M, and that two machines of type M are available. The two machines could be tooled identically to perform all four of these operations or some subset of the four. Alternatively, if time and tool capacity exists, grouping could be avoided by providing one set of tools to one of the machines.

Algorithm 4 is a sequential procedure as stated, but it could be modified to more rigorously examine the interactions between jobs. For example, once jobs have been found machine time and tool feasible, they could be placed on an intermediate list. From this intermediate list, jobs could be further evaluated two at a time (or more) to take advantage of common tooling requirements, or batch splitting.

3. SUMMARY AND FUTURE RESEARCH

This paper presents several heuristics for determining how machine tool magazines in a flexible manufacturing system can be loaded to meet simultaneous production requirements of a number of different part types. This loading problem is multicriteria in nature, and hence, no one of the heuristics introduced would likely meet the needs of all FMSs. Future research is needed to better define the variety and character of FMS loading objectives, how the loading problem links with the other four FMS production planning problems presented, and how loading and real time scheduling of parts on a system interact. It seems that detailed simulation of real systems could be used to help analyze these issues.

REFERENCES

1. Mohammed Berrada and Kathryn E. Stecke, "A Branch and Bound Approach for Machine Loading in Flexible Manufacturing Systems," Working Paper No. 329, Division of Research, Graduate School of Business Administration, The University of Michigan, Ann Arbor MI (April 1983).
2. D. S. Johnson, "Near Optimal Bin Packing Algorithm," Ph.D. dissertation, Mathematics Department, M.I.T., Cambridge MA (1973).
3. Kathryn E. Stecke, "Experimental Investigation of a Computerized Manufacturing System," Master's Thesis, School of Industrial Engineering, Purdue University, W. Lafayette IN (December 1977).
4. Kathryn E. Stecke, "Production Planning Problems for Flexible Manufacturing Systems," Ph.D. dissertation, Department of Industrial Engineering, Purdue University, W. Lafayette IN (August 1981).
5. Kathryn E. Stecke, "A Hierarchical Approach to Production Planning in Flexible Manufacturing Systems," in Proceedings, Twentieth Annual Allerton Conference on Communication, Control and Computing, Monticello IL (October 6-8, 1982).
6. Kathryn E. Stecke, "Formulation and Solution of Nonlinear Integer Production Planning Problems for Flexible Manufacturing Systems," Management Science, Vol. 29, No. 3, pp. 273-288 (March 1983).
7. Kathryn E. Stecke and Thomas L. Morin, "Optimality of Balancing Workloads in Flexible Manufacturing Systems," European Journal of Operational Research (1984), forthcoming.
8. Kathryn E. Stecke and James J. Solberg, "Loading and Control Policies for a Flexible Manufacturing System," International Journal of Production Research, Vol. 19, No. 5, pp. 481-490 (September-October 1981).
9. Kathryn E. Stecke and James J. Solberg, "The Optimality of Unbalanced Workloads and Machine Group Sizes for Flexible Manufacturing Systems," Working Paper No. 290, Division of Research, Graduate School of Business Administration, The University of Michigan, Ann Arbor MI (January 1982).

PERFORMANCE MODELING AND EVALUATION OF FLEXIBLE MANUFACTURING SYSTEMS USING SEMI-MARKOV APPROACH

M. ALAM[1], D. GUPTA[2], S.I. AHMAD[3] and A. RAOUF[4]
[1]School of Computer Science, Univ. of Windsor, Windsor, Ont., N9B 3P4, Canada
[2]Dept. of Management Sci., Univ. of Waterloo, Waterloo, Ont., N2L 3G1, Canada
[3]School of Computer Science, Univ. of Windsor, Windsor, Ont., N9B 3P4, Canada
[4]Systems Engg. Dept., Box 128, Univ. of Petroleum & Minerals, Saudi Arabia

ABSTRACT

This paper presents a Semi-Markov modeling technique for exact and approximate analysis of queueing networks often encountered while designing Flexible Manufacturing Systems (FMSs). In this approach, the states are observed only at the times of transitions. The mean residence time in each state is obtained using the probability distribution function of the conditional state occupancy times. Several appropriate states are lumped together in an effort to reduce the complexity of the original state description. The reduced model, though approximate, is much simpler to analyse. The symmetric pattern is then exploited to obtain a simple computational algorithm. Next, several performance measures are evaluated using the steady state probability distribution of the Semi-Markov process. Analytical and simulation results are compared for several examples and are found to be quite close. Computation times for an illustrative example using the above methods are compared. The approximate method is found to be very efficient.

INTRODUCTION

Computer integrated systems have become an essential part of modern industrial complexes. These systems are often very large as well as complex and therefore cannot be understood by heuristic approach alone. Design of such complex systems with high standards of performance requires use of a systematic approach and application of appropriate mathematical tools in order to predict the essential performance measures. Queueing network models have been found to be quite suitable for capturing the essential attributes of such systems.

This paper deals with some exact and approximate solution procedures for queueing networks with blocking due to finite queue capacities and limited resources. Such queueing networks are often encountered in the design of Flexible Manufacturing Systems (FMSs). In this work, emphasis has been put on developing computationally efficient, yet nearly exact solution procedures. The proposed model has also been applied to several examples of Flexible

Manufacturing Systems to illustrate its potential benefits in an application environment.

Queueing network analysis finds extensive use in the performance evaluation of computer systems (refs.9,20). However, simplified procedures are available for only a certain class of networks called 'the local balance networks' (refs.3,8,10,12). Recently, efforts have been made to develop simple and efficient solution procedures for networks not satisfying local balance. These attempts have been directed mainly towards performance evaluation of multiprocessor systems with shared buses (refs.13,16,17). These systems possess limited number of buses and only a restricted number of requests can be queued at the global memories at any time. The underlying queueing models are non-decomposable and hence do not possess an easy exact solution. Some of these approaches, used to derive computationally efficient algorithms are listed below:

i) Method of surrogate delays by Jacobsen and Lazowska (ref.13).

ii) Approximate Markov Models by Marsan and Gerla (ref.17).

iii) Norton's Decomposition Theorem (ref.7) by Kriz (ref.16).

iv) More recent methods e.g. (ref.26), (ref.27) and (ref.28).

Flexible Manufacturing Systems (FMSs) may consist of a single or a small number of machines serviced by robots to very complex set of multiple machines serviced by automated and programmable materials handling equipment. A detailed description of FMSs and several case studies are described in (ref.1). Some of the significant contributions in the field of analytical modeling of these systems can be found in (refs.4,5,18,22,23,24). A large number of these models, notably (ref.22) and (ref.5) rely on queueing network analysis with product form solutions. Yao and Buzacott (ref.25) have recently developed an iterative decomposition technique that can handle non-separable networks also.

In the models presented here, the machining cells are considered to be the primary resources and the Automated Guided Vehicles (AGVs) as the secondary resources. This classification is important because an AGV is necessary only to access any machining cell from the central store. The AGV is not needed after the job has been trasported from the central station to the desired machining cell and is free to serve the next job in the input queue. We make use of a Semi-Markov modeling approach which realistically captures the important characteristics of an FMS and is general enough to be valid for systems with different parameters. Based upon this model, we present efficient algorithms for exact and approximate analysis and demonstrate their easy implementation on computers. The results of several examples using the analytical models are compared with the results of simulation experiments using the SLAM (ref.13) processor.

FMS ARCHITECTURE AND ITS GENERIC DESCRIPTION

Flexible manufacturing systems (FMS) are aimed at providing a factory operation which is able to adjust in real time to changes in product design, production goals and scheduling, workstations, material handling and routing, etc. This operation should allow for simulation to relate performance to change, and provide feedback on the operation to assist in making production decisions. Examples of changes are product kind vs. product mix, production goal to maximize resource utilization vs. MRP "wanted" dates, scheduling parts to machines in a fixed sequence vs. random sequence based on certain conditions, workstations size vs. capabilities, material handling increase in buffering capacity vs. buffer locations, routing in a fixed or a variable pattern vs. delays. Simulation requirements include shop floor scheduling, traffice control, tool allocation and actual production processing as well as those parts of product life cycle which precede or follow shop floor operations.

The basic hardware elements of a manufacturing cell are machining or assembly workstations with robot for loading/unloading workpieces, programmable controllers, DNC machines, etc. supported by an automated material handling system.

A manufacturing system is driven by input stimuli from the market in the form of direct product demand, market conditions and feedback on production. The activities required may be broadly classified as (a) management including strategic planning, (b) design, (c) production planning, and (d) production operation. We may say that information from management level (a) flows to one of the lower leves (b,c,d). We would also notice the need for information to flow from b to c, c to d, etc. We would also notice the need for information flows to be two-way. For example, management sends some product information to Engineering Design and requests a response to proceed further. Design may send information to Production Planning and requests a response on production, and once an item is in production it also receives performance information. Each of the activities a to d would normally be divided into several sub-activities (components) with appropriate interconnections. While an exhaustive list of these sub-activities is unnecessary for the purpose of describing FMS architecture, it is useful to list a representative sample as follows:

1. Allocation of operations to workstations with associated tooling requirements. This includes consideration of such things as machines and tools to be considered vs. operations and their sequencing.
2. Material handling with respect to transportation of parts to be retrieved/stored vs. transfer/discharge of parts at

a workstation i.e. setting up parts on pallets for machining, pallet exchange, removal of finished parts for pallets, etc.

3. Production scheduling operations in addition to those already included in 1 and 2 above i.e. material planning, procurement and shipment.
4. Data collection on production activities.
5. Modeling production capabilities and evaluating alternatives.
6. Repair and maintenance scheduling, evaluation and inventory.
7. Initial product design and performation simulation of its production.
8. Control, monitoring and inspection of production operations, machine/tool failure and product quality.
9. Product testing.
10. Product distribution.
11. Product planning, market forecasting, costing and market evaluation.

Materials handling subsystem is responsible for delivery of materials to machining centres and subsequent removal of finished parts. It is important that this movement of material is synchronized to machining operations in a manner that maximizes the utilization of production resources. This factor is considered critical to system performance in manufacturing. It is this subsystem which is the subject of further investigation in this paper.

The study presented here, considers an FMS with several machining cells serviced by automated guided vehicles (AGVs) and robots. A schematic of the FMS showing flow of jobs through the system is shown in Fig. 1. Each of the machining cells can service several jobs and can perform a variety of different operations simultaneously on various jobs. The item flow and input to the system is controlled by a central server. Each job undergoes an arrival delay (service but no queueing) during which it travels back to the central station after a service at one of the machining cells and the central server determines the sequence of operations to be performed next on the job. The central server also inputs necessary data to the central computer in order to program the DNC machines to perform the desired machining operations. The arrival delay is assumed to be exponentially distributed with a mean $1/\lambda$ for each job. Since a variety of different jobs having different sequence of operations and parameters are likely to be encountered in an FMS environment, it is logical to assume an exponential distribution for arrival delay. Next each job requests service at one of the work stations on a first-come-first-served (FCFS) basis. If none of the machines where a job could commence its next sequence of operations is available, it waits in a finite capacity local storage queue for the work station at the first available location. In order to access a work station or an associated local store, each job must acquire one of the free AGVs. All jobs compete for AGVs on an FCFS basis. In case all the local stores are full and/or none of the AGVs is available, the queue at the central station is blocked. It is assumed that after each cycle of service, jobs return to the central station from where they can either request for further service or leave the system to be stored at the AS/R (Automated Storage/Retrieval) area. If a job leaves the system, we assume that another job from the AS/R area takes its place immediately, keeping the total number of jobs in the system constant; thus yielding a closed queueing system.

THE QUEUEING NETWORK MODEL

We consider an FMS having a set of work stations, numbered 1, 2, ..., M. The central station is indexed as '0' and has a common storage area C large enough to accommodate all the N jobs in the system. A schematic of its queueing netowrk with contention for simultaneous resource possession is shown in Fig. 2. Each work station has an associated local storage space. Some important characteristics of this system and the assumptions involved in deriving the various models are as follows:

(i) Each of the M machining cells has multiple machines. The ith machining centre can service m_i jobs at a time.

(ii) Each machining cell is an independent unit capable of performing a number of different machining operations.

(iii) There are a fixed number of jobs, N, circulating in the system at any time.

(iv) There are K AGVs available to the system.

(v) Each job undergoes an arrival delay with mean = $1/\lambda$. This is the time taken by the central server to program the DNC machines plus the time taken by jobs to travel back to the central station after completing a service cycle at one of the machining cells.

(vi) The AGV travel times are assumed to be deterministic with a mean = $1/\mu_t$. The choice of deterministic travel times is realistic and practical. The modeling technique presented here is capable of handling any general distribution with a known density function and having rational Laplace transform. Upon completing their task, the AGVs are released immediately (with zero delay) for next service.

(vii) There are R classes of jobs. Processing time of the ith class job at the jth machining cell could be described by any general distribution with mean = $1/\mu_{i,j}$ and probability density function $\gamma_i(t)$. $\gamma_i(t)$ is assumed to possess rational Laplace transform. We may assume that the grouping of machines is done in such a manner that all machining cells have nearly equal capabilities. Therefore, it will be logical to assume that the mean service time for all jobs of class i is $1/\mu_i$ for all machining cells. Thus the machining cells will be considered as homogeneous servers yielding better load balancing and increased productivity (ref.5).

(viii) A job can request service at any one of the machining cells. The probability with which the ith job requests the jth work station is $p_{i,j}$, $i = 1, 2, \ldots, N$ and $j = 1, 2, \ldots, M$; where,

$$\sum_{j=1}^{M} p_{i,j} = 1 \qquad \forall\ i = 1, 2, \ldots, N$$

If the jobs can request any machining cell arbitrarily and have no preferences for or biases against any machining cell then $p_{i,j} = 1/M$, $\forall\ i,j$.

(ix) At the ith machining cell, there is a local store that can hold Z_i jobs. This storage space is occupied only by those jobs whose next operation is at station i.

(x) The time required to bring a job back from machining cells to the central store need not be considered as the feedback flow is normally managed by some facility other than the AGVs such as a roller conveyor or a carousel. Moreover, the arrival delay takes care of the time spent before a job is ready to circulate through the system again.

For convenience an FMS with N jobs, M machining cells and Z local stores for all machines, will be denoted as an NxMxZ system. Note that this representation is used only when all the machining cells can process equal number of jobs simultaneously and have identical local stores.

THE SEMI-MARKOV MODEL

A Semi-Markov process is a random process whose successive state occupancies are governed by the transition probabilities of a Markov process, but whose stay in any state is described by a random variable that depends on the state presently occupied and on the state to which the next transition will be made (ref.11). The process is Markovian only at certain points belonging to the state space where transitions take place. It is easy to see that the queueing network described in section 3 has a state dependent arrival process and therefore, a product form solution does not hold. However, a Semi-Markov approach could be used to solve for the steady state probabilities of occupying any state of the stochastic process (refs.2,21).

Let,

$x_{i,j}$ = a random variable representing the waiting time in state i given that the next transition will be to state j.

$f_{i,j}(.)$ = probability density function of waiting times $x_{i,j}$'s.

$F_{i,j}(.)$ = distribution function of waiting time $x_{i,j}$.

$a_{i,j}$ = ij^{th} element of the transition probability matrix A of an associated Markov chain.

When the process has just entered state i, the next state j is selected according to the transition probability $a_{i,j}$, but once j has been selected, the waiting time $x_{i,j}$ is specified by the function $F_{i,j}(.)$. The unconditional mean waiting time in a state i, τ_i, can be obtained as:

$$\tau_i = \int_0^{\infty} \{ \sum_{j=1}^{n} a_{i,j}(t)\ f_{i,j}(t)\} \ .\ t\ dt \qquad (1)$$

where

n = total number of states in the state space.

The steady state probability vector of the associated Markov chain, $\P\ (= \P_1, \P_2, \ldots, \P_n)$, can be found by solving the following set of linear equations:

$$\P A = \P \qquad (2)$$

$$e'\P = 1 \qquad (3)$$

$$\P \geq 0 \qquad (4)$$

where e' is a row vector of 1's of appropriate size. Now, from the results established in (ref.2) and (ref.21), and fundamentals of matrix algebra, the steady state probabilities of the process can be found by:

$$P_i = \tau_i \P_i / (\sum_{j=1}^{n} \tau_j \P_j) \qquad (5)$$

Therefore, the problem of finding the steady state probability vector $P\ (= [P_i])$ is reduced to that of

(a) determining the state space in order to obtain the transition points of an imbedded Markov chain;

(b) finding the transition probability matrix $(a_{i,j})$ and unconditional mean waiting time in any state, τ_i, assuming that $f_{i,j}(.)$ are known;

(c) and finally, solving a system of linear equations (2), (3), (4) and (5) to obtain P_i's.

It is easy to see that the degree of complexity of this approach increases as the level of detail described by the state increases. Several performance measures can be found knowing the steady state probabilities P_i's, as explained in the following section.

PERFORMANCE MEASURES

Let $N_{i,r}$ denote the number of jobs of class r, being serviced when the system state is i. The expected number of jobs being machined in steady state is

$$E[N] = \sum_{i=1}^{n} \left(\sum_{r=1}^{R} N_{i,r} \right) P_i \qquad (6)$$

The throughput of a queueing network is defined as the rate at which the jobs are cycling through the system. The average utilisation of each server can be found by knowing the state description and the probability of the server being busy in the steady state. For the jth parallel server, average utilisation,

$$\Gamma_j = 1 - (\text{probability that the } j^{th} \text{ server is idle}) \qquad (7)$$

Throughput of the queueing network is then,

$$TH(N) = \sum_{j=1}^{M} \Gamma_j \mu_j \qquad (8)$$

This can also be computed using the Little's formula as follows:

$$TH(N) = \sum_{i=1}^{n} \left(\sum_{r=1}^{R} N_{i,r} \mu_r \right) P_i \qquad (9)$$

Now the average delay per job can be found as,

$$D = (N - E[N])/TH(N) \qquad (10)$$

In the present work, D is treated as the performance index of the CMS. Other indices like the idle time of servers, average queue lengths and resource utilisation, etc. can be found by using appropriate queueing formulae.

THE EXACT METHOD

The state of the stochastic process is defined as

$$S = \{S_1, \propto_1, S_2, \propto_2, \ldots, S_N, \propto_N\} \quad (11)$$

where

S_i = the state of the ith job, and

$\propto_i$ = the work station requested by job i.

State of the job can be one of the following:

$S_i = 0$ if the job is being machined at the $\propto_i{}^{th}$ station.

$S_i = j$ if the job is the jth one in the queue awaiting a free machine or an AGV at station $\propto_i$.

$S_i = -1$ if the job is being transported by the AGV to $\propto_i{}^{th}$ station.

Simalarly,

$\propto_i = 0$ if the ith job is at the central station.

$\propto_i = \{1,2,\ldots,M\}$ if the ith job requests for any of the M machining cells.

A transition occurs whenever a job changes its state. It is evident that the state space possesses Markov property at state transition instants. However, even for a small 5x2x1 FMS (5 jobs, 2 machines and unit local stores), with 2 AGVs, the number of states in the state space can be as high as 90! Therefore, it is impractical to work with this model.

The state space can be greatly reduced by lumping appropriate states. The following lumped state definition can be directly obtained from the general theory of lumpability (ref.15).

$$S = \{n_1, n_2, \ldots, n_M, q_0, q_1, q_2, \ldots, q_M, n_k\} \quad (12)$$

where

$n_i \in \{1, 2, \ldots, m_i\}$ and denotes the total number of jobs being machined at the ith machining station, ($\forall i = 1, 2, \ldots, M$)

$q_0 \in \{1,2, \ldots, N\}$ and is the total number of jobs waiting at the local store of machine i at any time, ($\forall i = 1, 2, \ldots, M$)

$n_k \in \{1, 2, \ldots, k\}$ and denotes the current number of free AGVs out of the total number k.

The process described by the above state description retains its Markov property at points of transition when the service requirements of different jobs types are identical. In other works, the lumping is exact only when the input jobs have a single job class. In the event that different job classes exist, this model can still be used as it satisfies the weak lumpability (ref.15) conditions when the starting probability vector renders equal probability for any one of the N jobs to be transported to a station i. In other words, if $P_{i,j} = 1/M \; \forall_{i,j}$, the weak lumpability conditions are satisfied. The transition rate is calculated as an average rate of service for all job classes.

$$\mu_{av} = 1/R\{\mu_1 + \mu_2 + \ldots \mu_R\}$$

A bench mark example of this model is solved in section 8 and the method of finding transition probabilities is explained in Appendix A.

THE APPROXIMATE METHODS

For the exact method of section 6, the state space grows rapidly with an increase in system size. Dealing with large matrix computations represented by (2), (3), (4) and (5) is both expensive and impractical. Moreover, no recursive form or inherent symmetric structure is observed in the state transition diagram to facilitate a reduction in computational work. Hence no algorithmic solution procedure can be established using the exact model.

The difficulty in above modeling procedure arises because of the detailed information that is contained in the state description. However, in many situations, we are only interested in the effect of variations in some design parameter e.g. number of AGVs, capacity of local stores, etc., on the overall system performance. Hence, by reducing the information content in the state definition, it is possible to develop approximate solution methods which are computationally very efficient. Two levels of state space reduction have been analysed in this work. The states of the reduced state space do not lie on a first order semi-Markov chain. However, the system behaviour is analysed under the assumption that the transitions between states will satisfy the Markov property. The expected results will not be exact and will have to be tested for accuracy.

Model #1

The state description for this model is,

$$S = \{n_m, q_0, q_1. q_2, \ldots, q_M, n_k\} \tag{13}$$

where

n_M = total number of jobs being machined at all the M machining cells, where $n_M \in \{1, 2, \ldots, \sum_{i=1}^{M} m_i\}$.

The remaining parameters are the same as in (12). Note that this model does not distinguish between jobs being machined at different work stations and is therefore exact only for homogeneous server case with a single job class. The lumping of states only satisfies conditions of weak lumpability with an initial vector that assigns equal probability to the system states with different jobs but with the same number at different machining locations and local stores. An example of an FMS has been solved with this method in section 8. The method of finding transition probabilities is explained in Appendix A.

Model #2

The state description for this model is,

$$S = \{n_m, q_0, q_M\} \tag{14}$$

where

$n_m \in \{1, 2, \ldots, \sum_{i=1}^{M}\}$ and denotes the total number of jobs being machined in the system.

$q_0 \in \{1, 2, \ldots, N\}$ and denotes the number of jobs queued at the central station '0' either because of nonavailability of a local storage space or an AGV.

$q_M \in \{1, 2, \ldots, \sum_{i=1}^{M} Z_i\}$ and denotes the total number of jobs queued at the local stores in the system.

The transition probabilities can be found by knowing the distribution functions of different types of services and delays as explained in Appendix A.

EXAMPLES AND RESULTS

An FMS with 2 machining cells having unit local stores and 2 AGVs is analysed as a benchmark example by various methods. The system is denoted as an Nx2x1 CMS with 2 AGVs. Each machining cell can service only one job at a time. Furthermore, we assume that,

μ_t = 0.5 jobs/time unit

μ_1 = $_2$ = 0.2 jobs/time unit

λ = 1 job/time unit

Moreover, the service time of the machining cells are assumed to be exponentially distributed for sake of simplicity. Figs. 3, 4 and 5 show state transition diagrams of the FMS example using the exact model and the approximate models #1 and 2 respectively. The operation of this FMS is simulated using the SLAM package (ref.19). Fig. 13 depicts the schematic diagram of a simulation network for the same example. System processing delay is computed by each method as the number of jobs in the system is varied from 5 t0 30. Table 1 gives the results of delay and average throughput rate for this example.

The results obtained by the exact analysis will always be within the 99% confidence interval of the results of simulation (see Fig. 6). The results of approximate analysis #1 and #2 are also very close as seen in figures 8 and 10. The approximation #1 usually gives upper bounds on system delay while approximation #2 gives lower bounds. The lower bounds can be explained intuitively because grouping several machining cells and local stores leads to a random redistribution of jobs to machining cells, and thus reduces congestion. The bounds are quite tight as the % errors are well below 10% in most of the cases.

The queueing network of figure 2 can be considered to consist of two load dependent subsystems. For instance, when the number of jobs in the system is low (light load condition) the delay caused by nonavailability of AGVs is a significant part of the system delay. But, during heavy load conditions, there are always some jobs waiting in the queue at the local stores and since for our example, $\mu_i < \mu_t$, the secondary resource subsystem delay does not significantly contribute towards system delay. Thus, by using the analytical techniques, we obtain asymptotic bounds on mean delay. The uppwer bounds are obtained by considering congestion due to AGV subsystem and the lower bounds by ignoring the delay due to AGVs. The simulation results are found to be within the asymptotic bounds as shown in figures 7, 9 and 11. The approximation is found to be very good when the number of jobs in the system is large.

System throughput obtained by the three methods and simulation experiments are plotted in Fig. 12 and found to be very close. Requirements of CPU time using an IBM 3031 computer for this example are typically 75 seconds for simulation, 8.4 seconds for the exact method and 7.5 and 5.5 seconds for approximation #1 and 2 respectively. The saving in CPU time are expected to be greater for larger systems.

CONCLUSIONS

Some exact and approximate Semi-Markov solution procedures have been presented and applied to the problems related to the FMSs, where Markov modeling approach fails. The approximate analysis has been found to be highly accurate and efficient for the chosen benchmark example. It has also been demonstrated that analytical methods use much less computer time compared to simulation models and are easy to use. Therefore, these models are of significant value to systems analysts and designers for studying the overall system performance.

The results of the approximate models presented in this paper are very close to those of exact model when the number of jobs in the system is large.

In actual practise, FMSs are designed to improve productivity and thus normally carry fairly heavy loads. The techniques developed here are likely to be very useful for most real life FMS environments. As far as small systems are concerned they could easily be analysed by the exact analysis.

The method developed in this paper offers complete flexiblity in the choices of system size and service time distributions provided they are known and possess rational Laplace transform.

The model presented here is currently being tested for a wide range of design variables and different distributions of machine service time.

APPENDIX A

First, consider the exact model with M machining cells, N jobs, m_i servers per machining cell, Z_i local stores and K AGVs. Let the state of the system at current time be

$$S^0 = \{n_1, n_2, \ldots, n_M, q_0, q_1, \ldots, q_M, n_k\} \tag{15}$$

Also let i denote the number of machining cells with at least one job being serviced and ℓ denote the number of machining cells with at least one machining or local store location free. In other words, M-ℓ machining centres and their associated buffers are full. Transitions can take place to at most (2M+1) neighbouring states from S^0. If a service takes place at one of the i machining centres, the resulting states are,

$$S_j^1 = \{n_1, n_2, \ldots, n_j, n_{j+1}, \ldots, n_M, q_0+1, q_1, q_2, \ldots, q_j-1, q_{j+1}, \ldots, q_M, n_k\} \tag{16}$$

$$\forall j,\ q_j \neq 0\ , \qquad j \in \{1, 2, \ldots, i\}, \text{ and } i \in \{0, 1, 2, \ldots, M\}$$

or

$$S_j^1 = \{n_1, n_2, \ldots, n_j-1, n_{j+1}, \ldots, n_M, q_0, q_1, q_2, \ldots, q_M, n_k\} \tag{17}$$

$$\forall_j,\ q_j = 0\ , \qquad j \in \{1, 2, \ldots, i\} \text{ and } i \in \{0, 1, 2, \ldots, M\}$$

Similarly, if n ≠ 0, $q_0 \neq 0$ and one of the jobs at the central store acquires an AGV, a transition will take place to,

$$S^2 = \{n_1, m_2, \ldots, n_M, q_0-1, q_1, q_2, \ldots, q_M, n_k-1\} \tag{18}$$

Another possibility is that a service is completed by an AGV and a job is deposited at one of the free machining locations or their associated local stores. The resulting states are,

$$S_j^3 = \{n_1, n_2, \ldots, n_j+1, n_{j+1}, \ldots, n_m, q_0, q_1, q_2, \ldots, q_M, n_k+1\} \tag{19}$$

$$\text{if } q_j = 0 \text{ , } n_j < m_j \text{ and } \forall_j \ \varepsilon \ \{1, 2, \ldots, \ell\} \text{ and } \ell \ \varepsilon \ \{0, 1, 2, \ldots, M\}$$

or

$$S_j^3 = n_1, n_2, \ldots, n_j, n_{j+1}, \ldots, n_M, q_0, q_1, \ldots, \ell_j+1, q_{j+1}, \ldots, q_M, n_k+1 \tag{20}$$

$$\text{if } q_j \neq 0 \text{ , } n_j = m_j \text{ , } \forall_j \ \varepsilon \ \{1, 2, \ldots, 1\} \text{ and } \ell \ \varepsilon \ \{0, 1, 2, \ldots, M\}$$

Having established the destination states of a transition, we need to know the distribution function of arrival delay, AGV service times, and machining times in order to find the transition probabilities. Assuming exponential distribution function for arrival delay and machining times, and deterministic AGV travel times, we obtain;

$$a_{S^0 \to S_j^1} = \frac{\mu_j[1 - \exp(-(\mu_1 + \mu_2 + \ldots \mu_i + q_0 \lambda)/n_k\mu_t)]}{(\mu_1 + \mu_2 + \ldots, \mu_i + q_0 \lambda)} \quad \forall i,j \tag{21}$$

Similarly,

$$a_{S^0 \to S^2} = \frac{q_0 \lambda [1 - \exp(-(\mu_1 + \mu_2 + \ldots, \mu_i + q_0 \lambda)/n_k\mu_t]}{(\mu_1 + \mu_2 + \ldots, \mu_i + q_0 \lambda)} \quad \forall i,j \tag{22}$$

and

$$a_{S^0 \to S_j^3} = 1/\ell[\exp(-(\mu_1 + \mu_2 + \ldots, \mu_i + q_0 \lambda)/n_k\mu_t)] \quad \forall i,j \tag{23}$$

Using the above results, the transition probability matrix could be constructed for any general FMS. Next, mean residence times in each state, τ_i's, could be calculated from (1). For example,

$$\tau_{S^0} = \int_0^{1/n_k{}^+_t} \{(\mu_1 + \mu_2 + \ldots, \mu_i + q_0 \lambda)[\exp(-(\mu_1 + \mu_2, \mu_i + q_0 \lambda)t)]$$

$$+ \exp(-(\mu_1 + \mu_2 + \ldots, \mu_i + q_0 \lambda)/n_k\mu_t).\delta(t - 1/n_k\mu_t)\} \, . \, t \, dt \qquad (24)$$

where

$\delta(t - 1/n_{k\,t})$ is the density function of a unit impulse function, and

$$\int_0^x \delta(t - 1/n_k\mu_t)dt = \begin{cases} 1 & \text{if } x \geq 1/n_k\mu_t \\ 0 & \text{otherwise} \end{cases} \qquad (25)$$

Using (21), (22), (23), (24), (25) and the results established earlier, steady state probabilities of the stochastic process underlying the FMS operation could be evaluated.

Next, we consider the approximate analyses for an FMS with same parameters as described above.

Approximation #1

Once again let the current state of the system be S^0.

$$S^0 = \{n_M, q_0, q_1, q_2, \ldots, q_M, n_k\} \qquad (26)$$

The states to which transitions may take place are,

$$S_j^0 = \{n_M-1, q_0+1, q_1, \ldots, q_M, n_k\} \text{ and } q_j = 0, \; j \in \{1, 2, \ldots, i\} \qquad (27)$$

or

$$S_j^1 = \{n_M, q_0+1, q_1, \ldots, q_j-1, q_{j+1}, \ldots, q_M, n_k\}$$

$$\text{if } q_j \neq 0, \; j \in \{1, 2, \ldots, i\} \qquad (28)$$

and,

$$S^2 = \{n_M, q_0-1, q_1, q_2, \ldots, q_M, n_k-1\} \qquad \text{if } n_k > 0 \qquad (29)$$

$$S_j^3 = \{n_M+1, q_0, q_1, q_2, \ldots, q_M, n_k+1\}$$

$$\text{if } q_j = 0, \text{ or } n_M < \sum_{i=1}^{M} m_i \;\&\; n_k < k \qquad (30)$$

or

$$S_j = \{n_M, q_0, q_1, \ldots, q_j+1, q_{j+1}, \ldots, q_M, n_{k+1}\}$$

$$\text{if } q_j \neq 0, \text{ or } n_M = \sum_{i=1}^{M} m_i \;\&\; n_k < k \quad \text{where} \quad j \in \{1, 2, \ldots, \ell\} \tag{31}$$

Transition probabilities can now be obtained in the same way as for the case of exact analysis.

$$a_{S^0 \to S_j^1} = \frac{(\mu_1 + \mu_2 + \ldots, \mu_i)}{i} \quad \frac{[1-\exp(-(\mu_1+\mu_2+\mu_3\ldots\mu_i+q_0\lambda)/n_k\mu_t)]}{(\mu_1+\mu_2+\ldots\mu_i+q_0\lambda)} \quad \forall i,j \tag{32}$$

$$a_{S^0 \to S^2} = \frac{(q_0\;)[1-\exp(-\mu_1+\mu_2+\ldots\mu_i+q_0\lambda)/n_k\mu_t)]}{(\mu_1+\mu_2\;+\ldots\mu_i+q_0\lambda)} \quad \forall i,j \tag{33}$$

$$a_{S^0 \to S_j^3} = 1/\ell \; \exp(-(\mu_1+\mu_2+\ldots\mu_i+q_0\lambda)/n_k\mu t) \quad \forall i,\ell \tag{34}$$

Approximation #2

Let the current FMS state be S^0 such that,

$$S^0 = \{n_M, q_0, q_M\} \tag{35}$$

For this model, the number of free AGVs at any time, $n_k = K - [N-(n_M+q_0+q_M)]$. The states to which transitions may take place are,

$$S^1 = \{n_M-1, q_0, q_M\} \qquad \text{when} \qquad n_m > 0 \tag{36}$$

or

$$S^1 = \{n_M, q_0, q_M-1\} \qquad \text{when} \qquad q_M > 0 \tag{37}$$

and

$$S = \{n_M, q_0-1, q_M\} \qquad \text{when} \qquad n_k > 0$$

$$S^3 = \begin{cases} \{n_M+1,\ q_0,\ q_M\} & \text{if } n_M < \sum_{i=1}^{M} m_i \quad (38) \\ \{n_M,\ q_0,\ q_M+1\} & \text{if } q_M < \sum_{i=1}^{I} Z_i \quad (39) \end{cases}$$

where

$$I = \text{largest integer} \leq \frac{n_M}{m_i}$$

Transition probabilities are,

$$a_{S^0 \rightarrow S^1} = \frac{n_{M\ av.}[1-\exp(-(\mu_M\mu_{av}+q_0\lambda)/n_k\mu_t]}{(n_M\mu_{av}+q_0\lambda)} \qquad (40)$$

where

$n_k = K - [N - (n_m + q_0 + q_M)]$ and ${}_{av} = 1/R[\mu_1 + \mu_2 + \ldots, \mu_R]$

$$a_{S^0 \rightarrow S^2} = \frac{q_0\ [1-\exp(-(n_M\mu_{av}+q_0\lambda)/n_k\mu_t)]}{(n_M\mu_{av}+q_0\lambda)} \qquad (41)$$

and

$$a_{S^0 \rightarrow S^3} = \exp(-(n_M\mu_{av}+q_0\lambda)/n_k\mu_t) \qquad (42)$$

The unconditional mean residence times can now be easily calculated using (1) and the transition probabilities.

REFERENCES

1 Anonymous, "Flexible Manufacturing Systems - Their Tremendous Potential - Handling's Critical Role - Seven of the World's Best", Mod. Matls. Handl., 51-72, (Sept. 7, 1982).

2 Barlow, R.E. and Proschan, F., "Mathematical Theory of Reliability", John Wiley & Sons, Inc., New York, 119-161, (1965).

3 Baskett, F., Chandy, K.M., Munitz, R.R. and Palacios, F., "Open, Closed and Mixed Networks of Queues With Different Classes of Customers", J.ACM, 22, 2, 248-260, (1975).

4 Buzacott, J.A., "Optimal Operating Rules for Automated Manufacturing Systems", IEEE Trans. on Automatic Control, AC-27, 1, 80-86, (1982).

5 Buzacott, J.A. and Shanthikumar, J.G., "Models for Understanding Flexible Manufacturing Systems", AIIE Trans., 12, 4, 339-350, (1980).

6 Buzen, J.P., "Computational Algorithms for Closed Queueing Networks with Exponential Servers", Comm. ACM, 16, 9, 527-631, (1973).

7 Chandy, K.M., Herzog, U. and Woo, L., "Parameteric Analysis of Queueing Networks", IBM J. Res. Develop., 19, 1, 36-42, (1975).

8 Chandy, K.M. Howard, J.H. and Towsley, D.F., "Product Form and Local Balance in Queueing Networks", J. ACM, 24, 2, 250-268, (1977).

9 Chandy, K.M. and Sauer, C.H., "Approximate Methods for Analysis of Queueing Networks Models of Computer Systems", Computing Surveys, 10, 8, 263-280, (1978).

10 Gordon, W.J. and Newell, G.F., "Closed Queueing Systems with Exponential Servers", Opns. Res., 15, 254-265, (1967).

11 Howard, R.A., "Dynamic Probabilistic Systems", Vol. II: Semi-Markov and Decision Processes, John Wiley, New York, (1971).

12 Jackson, J.R., "Network of Waiting Lines", Opns. Res., 5, 518-521, (1975).

13 Jacobson, R.A. and Lazowska, E.D., "Analysing Queueing Network with Simultaneous Resource Possession", Comm. ACM, 25, 142-151, (1982).

14 Keller, T.W., Jr., "Computer System Models with Passive Resources", Unpublished Ph.D. Dissertation, Univ. of Texas at Austin, Texas, (1976).

15 Kemeni, J.G. and Snell, J.L. "Finite Markov Chains", Van Nostrand, New York, (1960).

16 Kriz, J., "A Queueing Analysis of a Symmetric Multiprocessor with Shared Memories and Buses", IEE Proc., 30, Pt.E., 2, 83-89, (1983).

17 Marsan, M.A. and Gerla, M., "Markov Models for Multiple Bus Multiprocessor Systems", IEEE Trans., Comput., C-31, 239-248, (1982).

18 Nof. S.Y., Barash, M.M. and Solberg, J.J., "Operational Control of Item Flow in a Versatile Manufacturing System", Int. J. Prod. Res., 17, 5, 479-489, (1979).

19 Pritsker, A.A.B. and Pegden, C.D., "Introduction to Simulation and SLAM", Halsted Press, John Wiley, New York, (1979).

20 Sauer, C.H. and Chandy, K.M., "Computer Systems Performance Modeling", Prentice-Hall Inc., New Jersey, (1981).

21 Singh, C. and Billington, R., "System Reliability - Modeling and Evaluation", Hutchinson & Co. Ltd., U.K., (1977).

22 Solberg, J.J., "A Mathematical Model of Computerised Manufacturing Systems", Presented at the 4th Int. Conf. Prod. Res., Tokyo, Japan, (Aug. 1977).

23 Stecke, K.E. and Solberg, J.J., "Loading and Control Policies for a Flexible Manufacturing System", Int. J. Prod. Res., 19,5, 482-490, (1981).

24 Stecke, K.E., "Formulation and Solution of Nonlinear Integer Production Planning Problems for Flexible Manufacturing Systems", Mgnt. Sc., 29, 3, 273-282, (1983).

25 Yao, D.D.W. and Buzacott, J.A., "Closed Queueing Network Models of Flexible Manufacturing Systems", Presented at the Int. Conf. Prod. Res., Windsor, Ontario, Canada, (Aug. 1983).

26 Reiser, M., "Mean Value Analysis of Queueing Networks: A New Look at an Old Problem", Performance of Computer Systems, M. Arato et. al. (Eds.), North-Holland, 1979.

27 Suri, R., "New Techniques for Modelling and Control of Flexible Manufacturing Systems", IFAC Proc. 18th Triennial World Congress, Kyoto, Japan, Aug. 1981.

28 Suri, R. and Cao, X., "Optimization of Flexible Manufacturing Systems Using New Techniques in Discrete Event Systems", Procs. 20th Allerton Conf. on Communication, Control and Computing, Monticello, Illinois, Oct. 1982.

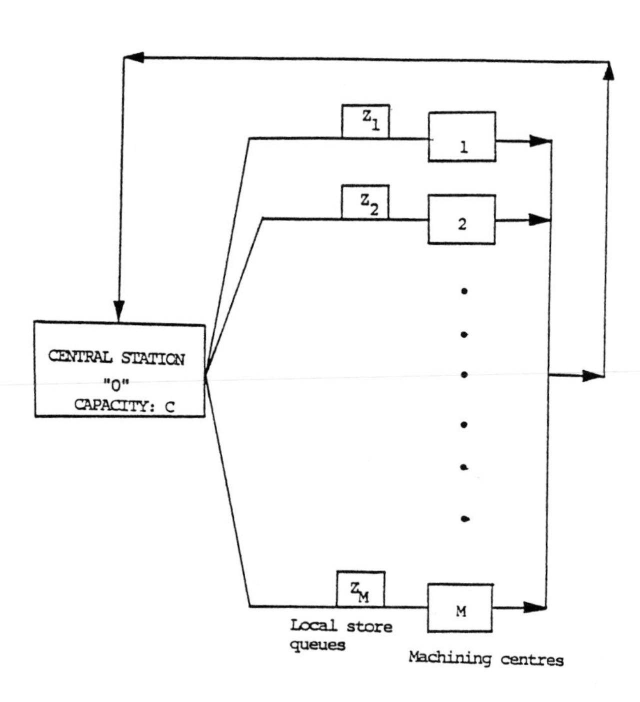

Figure 1: Schematic diagram of the central server model of an FMS showing item flow.

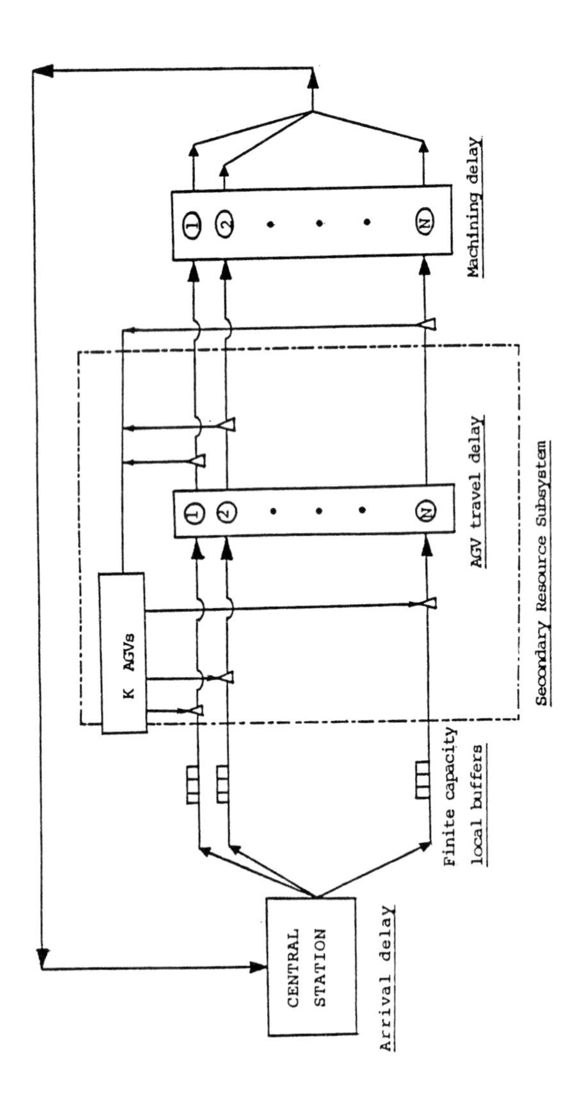

Figure 2: Queueing network model of the central station FMS with contention for AGVs: The Secondary Resource.

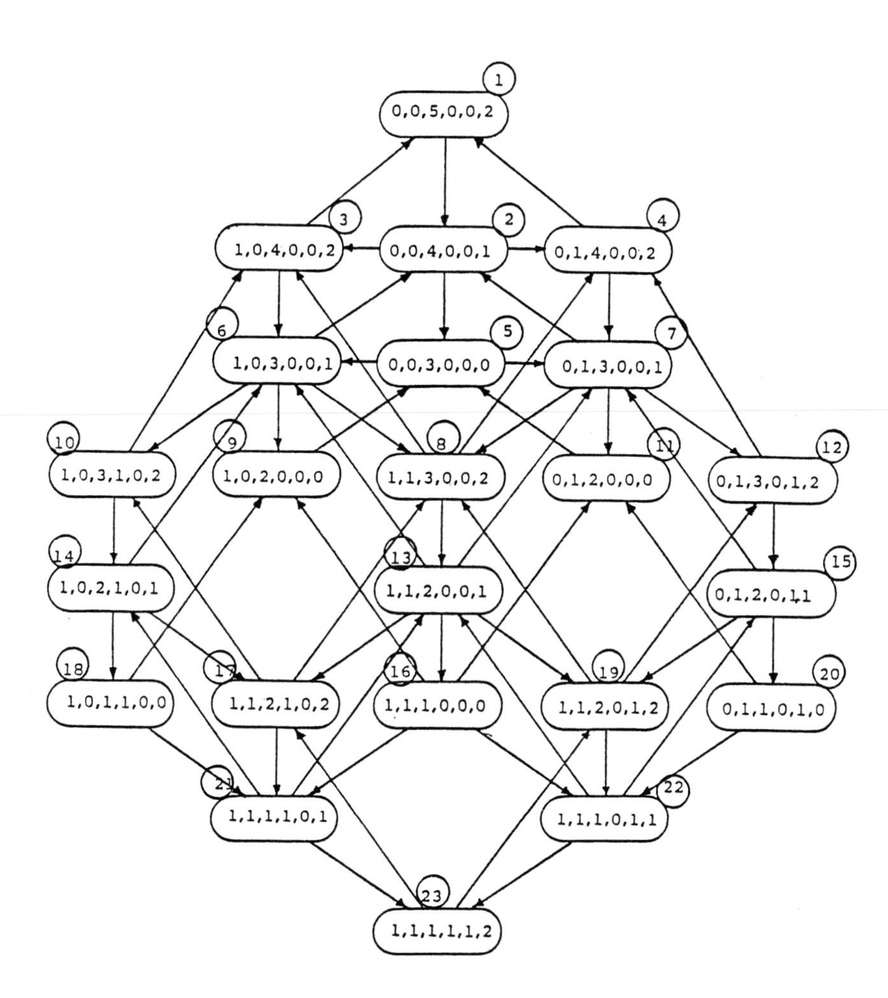

Figure 3: State transition diagram for a 5x2x1 FMS using the exact analytic model.

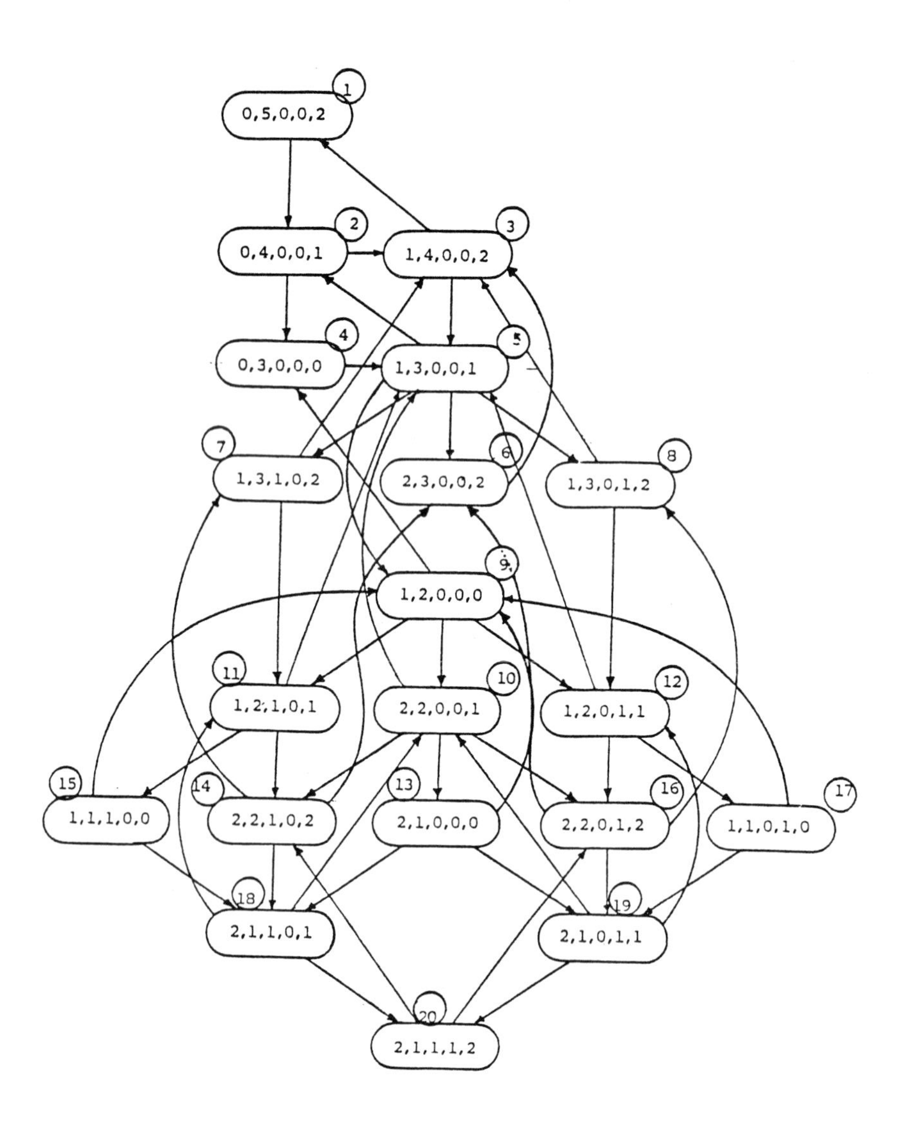

Figure 4: State transition diagram for a 5x2x1 FMS using the approximate model #1.

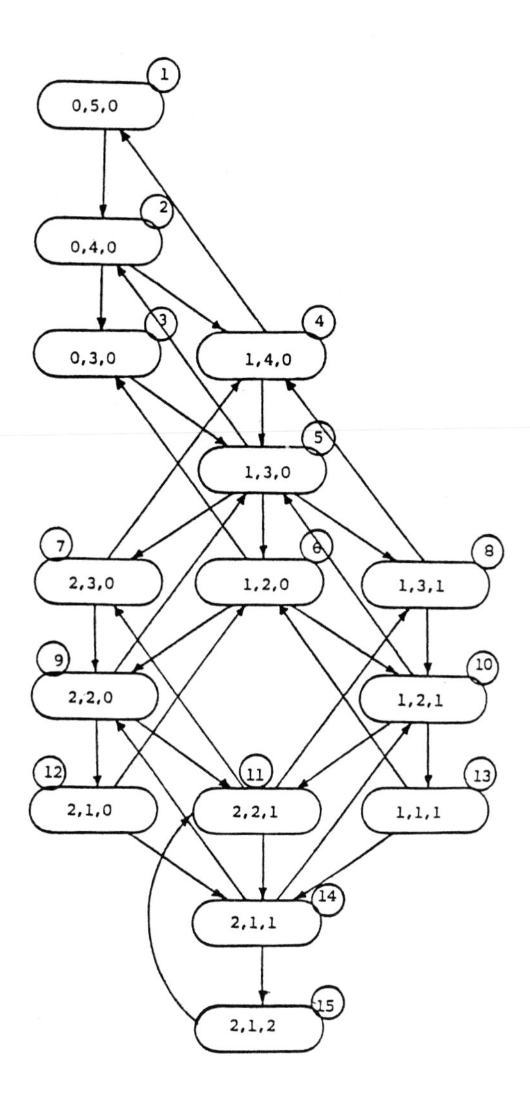

Figure 5: State transition diagram for a 5x2x1 FMS using the approximate model #2.

PERFORMANCE MEASURES OF A Nx2x1 FMS

WITH 2 AGVs - APPROXIMATE MODEL #1

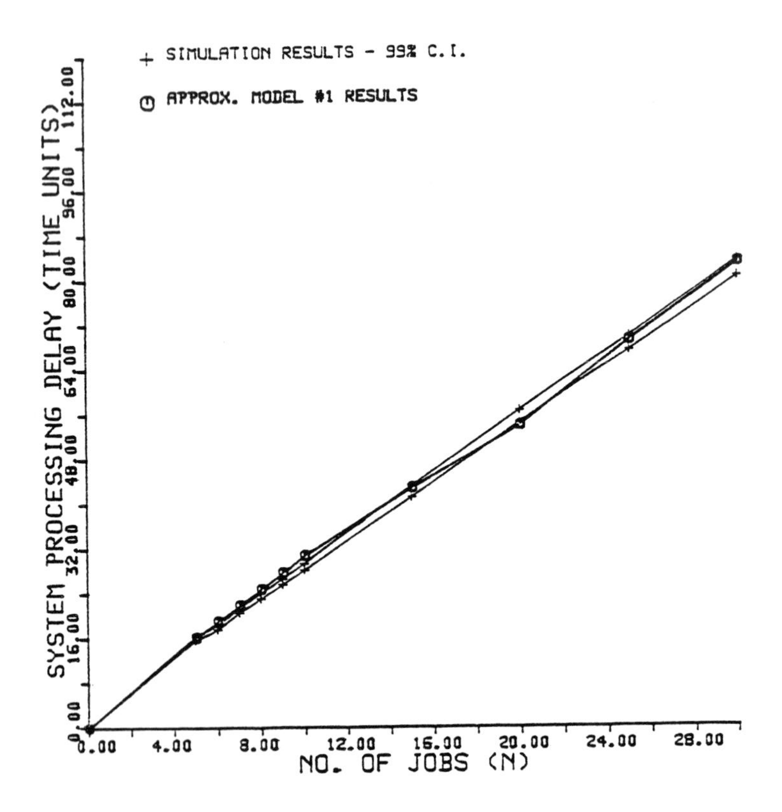

Figure 6: 99% confidence interval on results of simulation & results of approximate model #1.

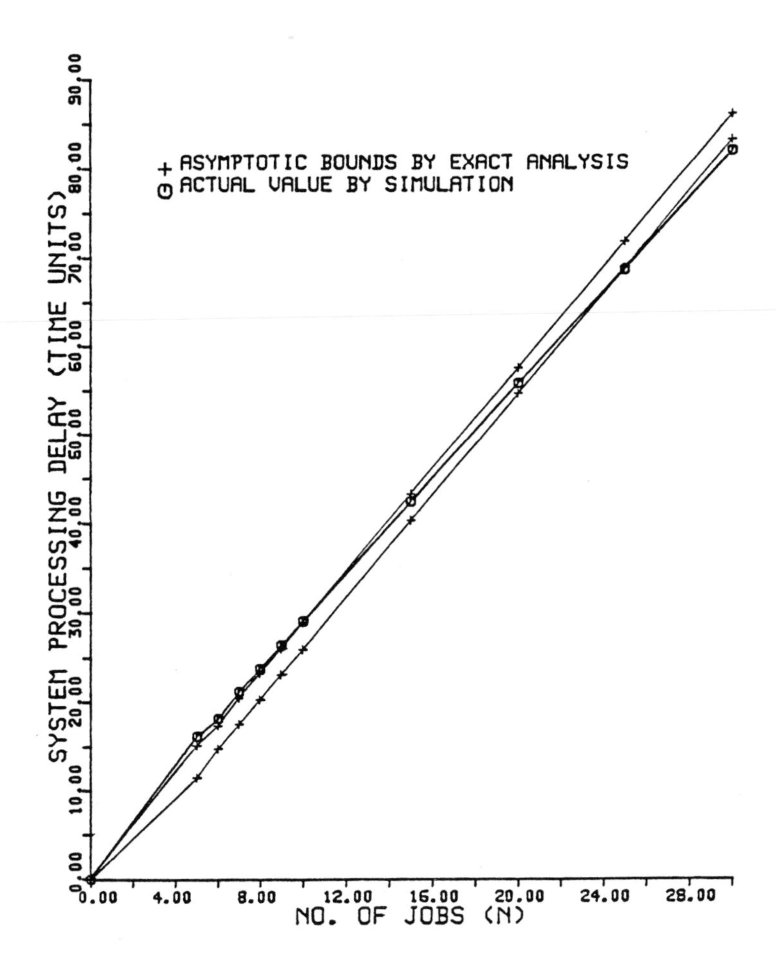

Figure 7: Asymptotic bounds on system delay using exact model.

PERFORMANCE MEASURES OF A Nx2x1 FMS

WITH 2 AGVs - THE EXACT MODEL

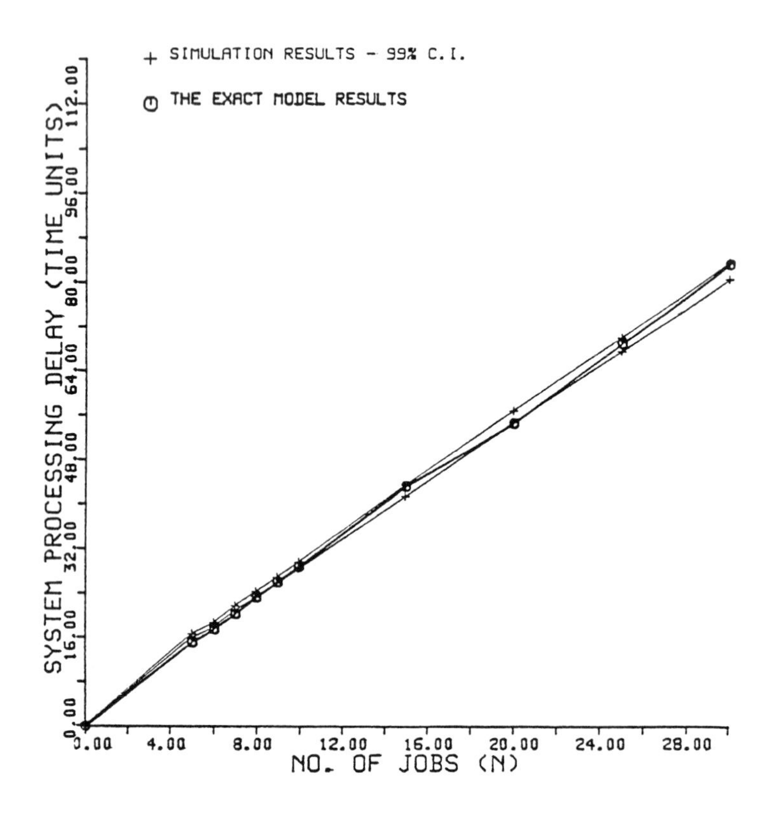

Figure 8: 99% confidence interval on results of simulation & results of the exact model.

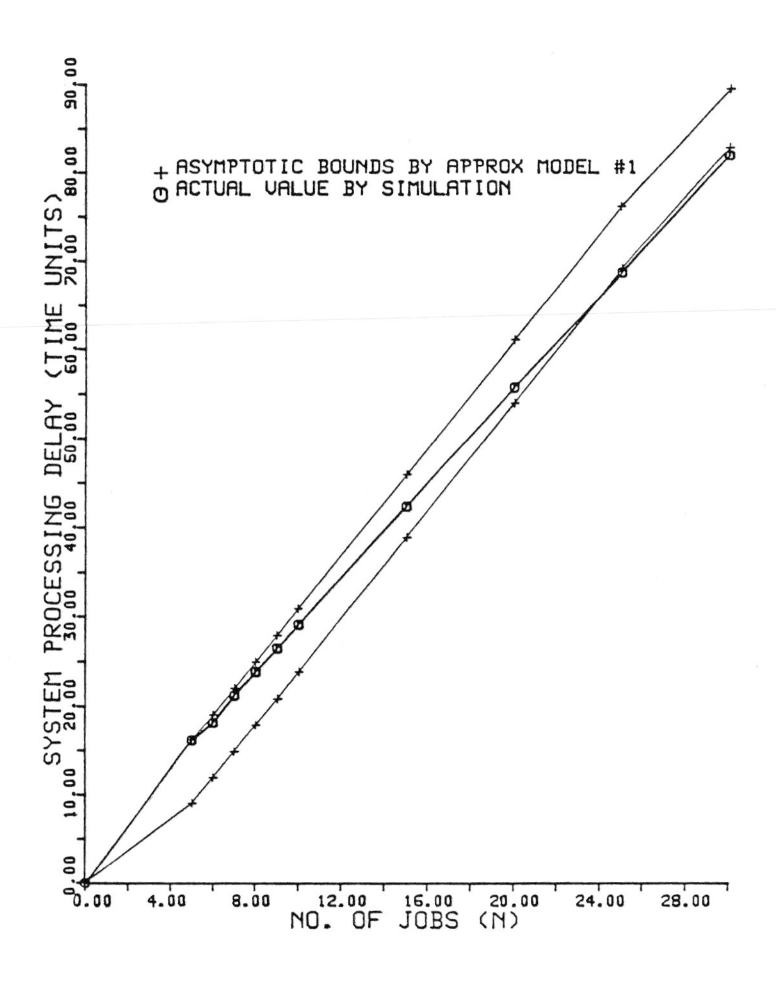

Figure 9: Asymptotic bounds on system delay using approximate model #1.

PERFORMANCE MEASURES OF A Nx2x1 FMS

WITH 2 AGVs - APPROXIMATE MODEL #2

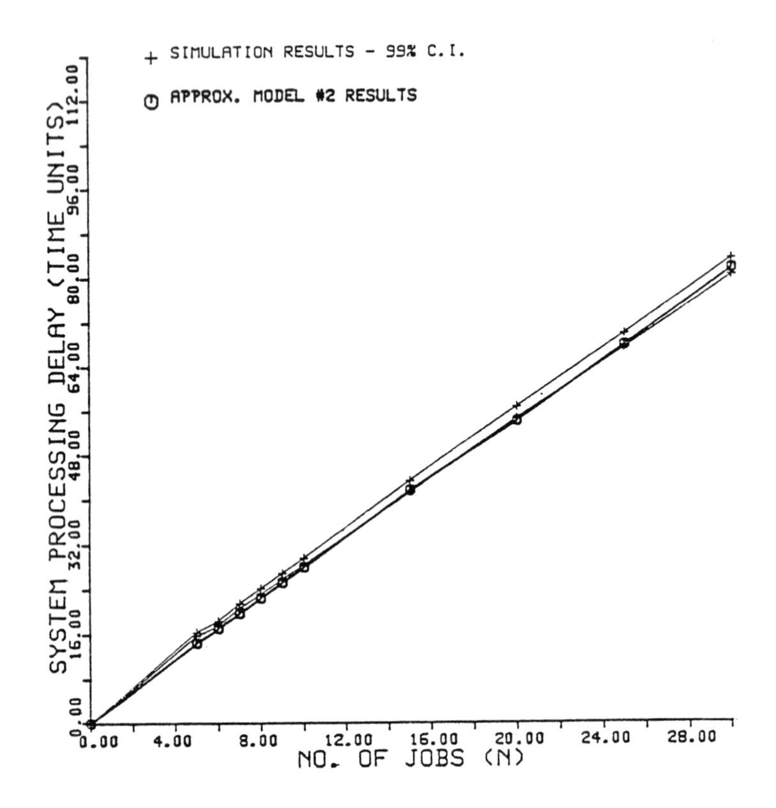

Figure 10: 99% confidence interval on results of simulation & results of approximate model #2.

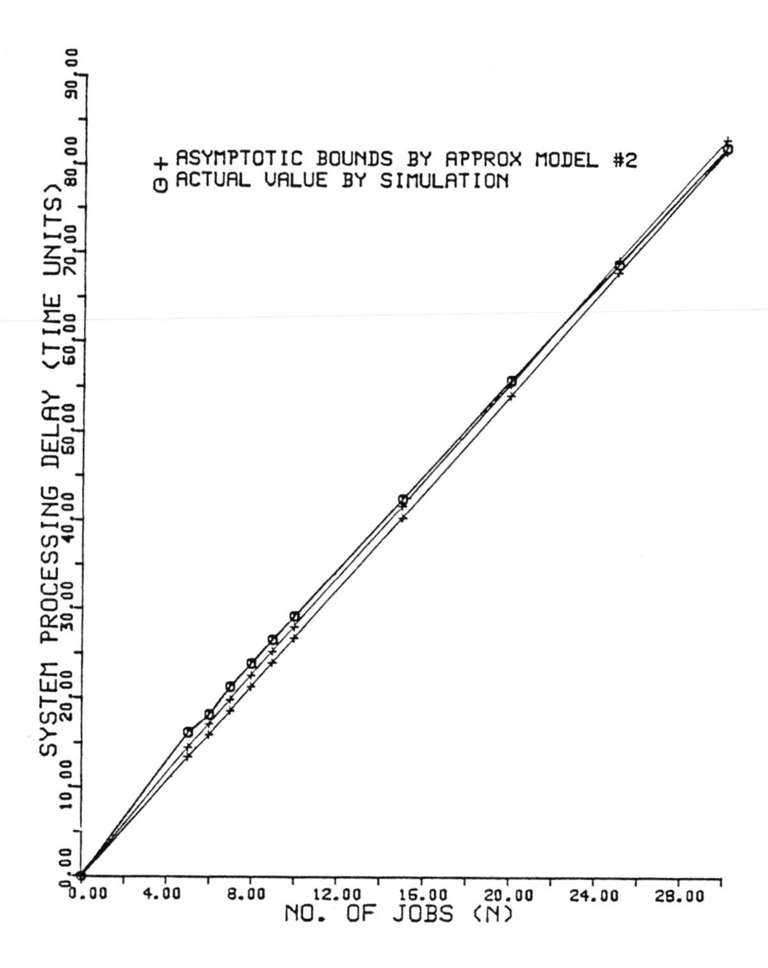

Figure 11: Asymptotic bounds on system delay using approximate model #2.

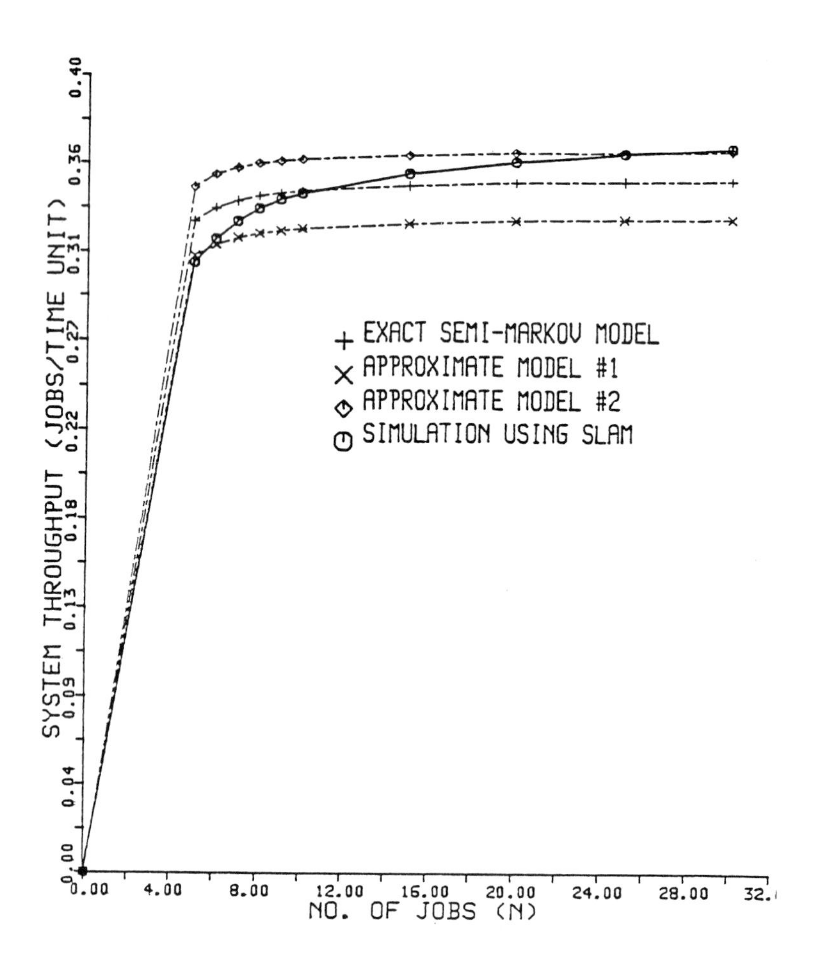

Figure 12: System throughput as a factor of no. of jobs.

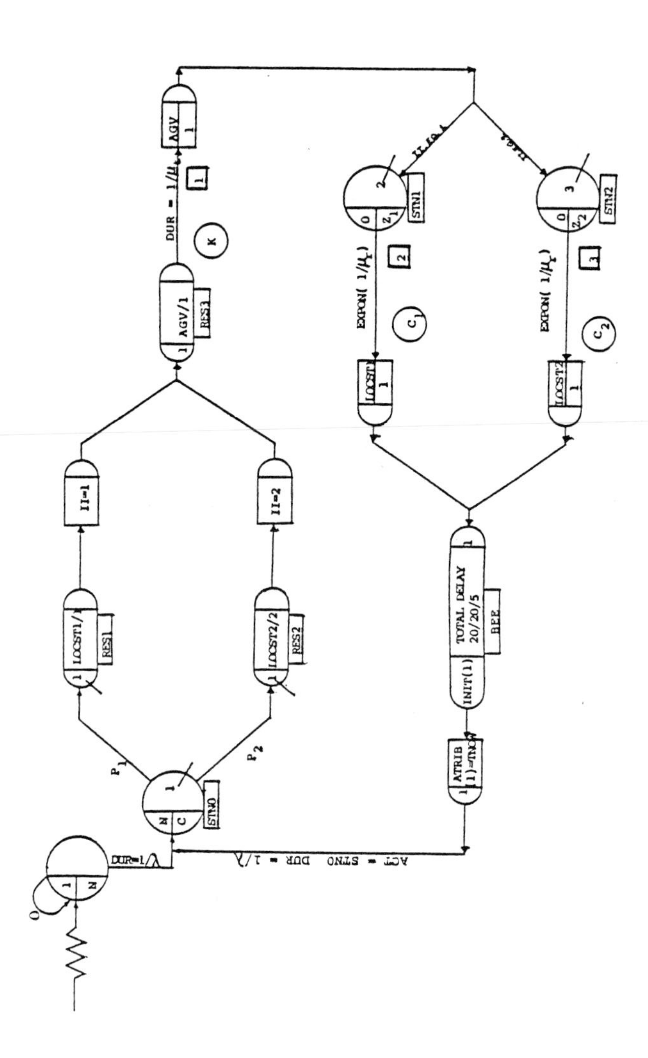

Figure 13: Schematic of the SLAM network for a 5x2x1 FMS example. (No. of AGVs = 2).

TABLE 1

System processing delay and throughput rate as a factor of number of jobs for an Nx2x1 FMS with 2 AGVs using simulation, the exact model and the approximations 1 & 2.

S. No.	No. of Jobs (N)	SYSTEM DELAY (TIME UNITS)				THROUGHPUT TH(N) JOBS/UNITS			
		Simulation 99% Confidence Interval	Exact Method	Approximate method #1	Approximate method #2	Simulat-ion	Exact Method	Approximate Method #1	Approximate Method #2
1	5	16.21 ± 0.331	15.154	16.24	14.57	0.30578	0.326090	0.308822	0.343148
2	6	18.21 ± 0.37	17.41	19.08	17.17	0.3173	0.332587	0.314447	0.349356
3	7	21.26 ± 0.412	20.562	22.00	19.84	0.32624	0.336289	0.317750	0.352739
4	8	23.84 ± 0.455	23.353	24.95	22.547	0.3326	0.338637	0.319892	0.354810
5	9	26.49 ± 0.545	26.169	27.94	25.267	0.33726	0.340248	0.321387	0.356195
6	10	29.13 ± 0.545	29.00	30.93	27.996	0.34038	0.341421	0.322489	0.357182
7	15	42.44 ± 0.78	43.242	42.928	41.71	0.35076	0.344428	0.325378	0.359680
8	20	54.78 ± 1.02	54.644	54.062	54.141	0.35636	0.345696	0.226628	0.360624
9	25	67.77 ± 1.25	68.963	69.203	67.895	0.36078	0.346395	0.327326	0.361161
10	30	81.05 ± 1.46	83.28	83.51	81.657	0.3635	0.346838	0.327772	0.361496

LOADING MODELS IN FLEXIBLE MANUFACTURING SYSTEMS

A. KUSIAK
Dept. of Industrial Engineering, Technical University of Nova Scotia,
P.O. Box 1000, Halifax, Nova Scotia, B3J 2X4 (Canada)

ABSTRACT

In this paper two different approaches to planning and scheduling in the classical manufacturing systems are overviewed. Based on these approaches, a planning and scheduling methodology for Flexible Manufacturing Systems (FMSs) is presented. One of the most important problems in the FMS methodology, the machine loading problem, is discussed to a great extent. Four models of the loading problem are formulated. Some of the algorithms for solving the loading problems are also discussed.

INTRODUCTION

Flexible Manufacturing Systems (FMSs) have gained a large number of applications. To date hundreds of these systems have been implemented around the world. Japan is leading in the FMSs international competition in terms of the number of applications and associated management and organization successes. It would be difficult to further develop FMSs without knowing their nature.

The aim of this paper is to discuss a new FMS planning and scheduling methodology. Based upon the framework of this methodology, a number of loading models are formulated.

Planning and scheduling problems are very difficult to solve computationally (see Lenstra 1977, VanWassenhove and DeBodt 1980). This is probably the main reason why a relatively small number of planning and scheduling theoretical results have been applied in practice. Taking into account on one hand the limitation of theory, and on the other the need for optimization of planning and scheduling decisions in FMSs, perhaps the best remedy is to reduce the size of these problems. This can be done by aggregation and decomposition.

It is not intended in this paper to survey all the planning and scheduling models encountered in the classical manufacturing systems, but to show only typical approaches. Readers interested in the classical planning and scheduling models may refer, for example, to the excellent surveys written by Gelders and VanWassenhove (1981), Graves (1981), Elmaghraby (1978), and Silver (1981).

PLANNING AND SCHEDULING IN THE CLASSICAL MANUFACTURING SYSTEMS

There are two basic approaches to production planning and scheduling in the classical manufacturing systems:

1. material requirements to planning (MRP)
2. hierarchical production planning (HPP).

Material Requirement Planning

To date, Material Requirement Planning (MRP) systems have been widely applied in industry. We should talk about a class of MRP systems rather than a single MRP approach, because of their diversity (Graves 1981). A typical MRP system is presented in Figure 1.

Based on product requirements which are either deterministic or forecasted orders, a master schedule is generated. The master schedule specifies the number of products to be manufactured within each time period (i.e. 1 month) over a planning horizon (i.e. 1 year). For each time period, products are exploded to parts. Of course, different products may consist of common parts, which are summarized. Production expressed in parts is split into batches. The formulae currently being applied for calculating the batch sizes are diverse. In general, they try to incorporate inventory holding costs and set-up costs.

Part batches are then assigned (loaded) to the most appropriate group of machines, then scheduled to ensure high machine rate utilization and satisfaction of the imposed constraints, i.e. due dates. The decision processes behind the MRP system are of an iterative nature. They are usually repeated many times until acceptable machine loads and sequence of batches are generated.

Hierarchical Production Planning

Hierarchical production planning systems take advantage of a natural approach for solving complex problems. Firstly, the production planning is solved on aggregated data over a long planning horizon. As a result of this, the entire production planning problem is decomposed into subproblems which are computationally easier to handle. Depending on the nature of the overall production planning problem, there may be two or more decision levels in the structure of the hierarchical planning system.

There are the following two advantages to an aggregate approach as opposed to a detailed planning approach (Bitram and Hax 1977):

- aggregate demands can be forecast more accurately than their disaggregate components
- there is a reduction in a problem size and degree of data detail.

The hierarchical methodology presented in this paragraph is based on the deterministic approach presented by Bitran, Haas and Hax (1981). Dempster (1982), and Dempster et al. (1981) have discussed a stochastic approach to the HPP sys-

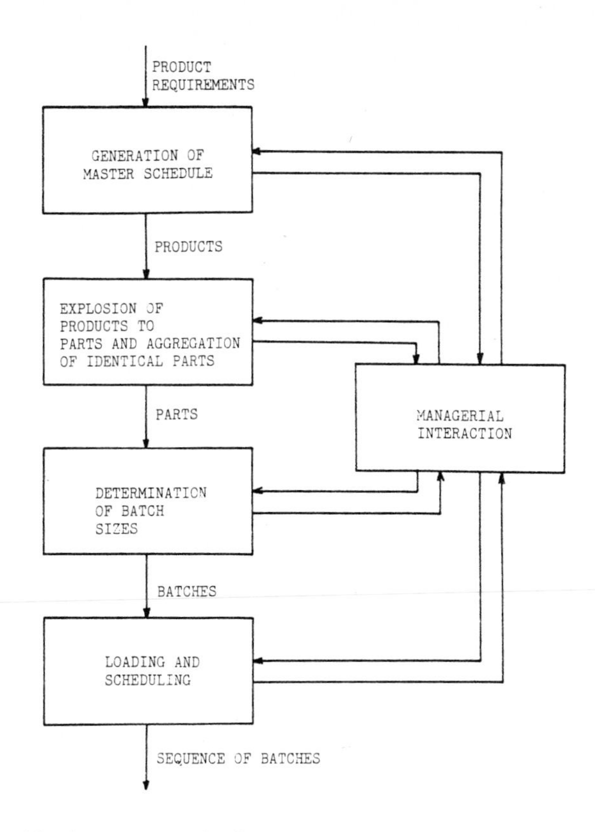

Fig. 1 Decision blocks in a typical MRP system

tems.

Before the hierarchical production planning approach is outlined the following terms are defined (see Bitran et al. 1981):

- product - an item delivered to the customer
- product type - a group of products having similar production costs and inventory holding costs
- product family - a group of items pertaining to the same product type and sharing similar setups.

The structure of the hierarchical production planning system is shown in Figure 2.

Mathematical formulations of each of the three decision blocks in Figure 2 are discussed in Bitran et al. (1981), Bitran and Hax (1977), and Bitran et al. (1982).

PLANNING AND SCHEDULING METHODOLOGY APPLICABLE TO FMSs

Most literature on FMSs has been published in the last decade, and there have not been many successful applications of planning and scheduling methodologies in the classical manufacturing systems; therefore, it would be difficult to

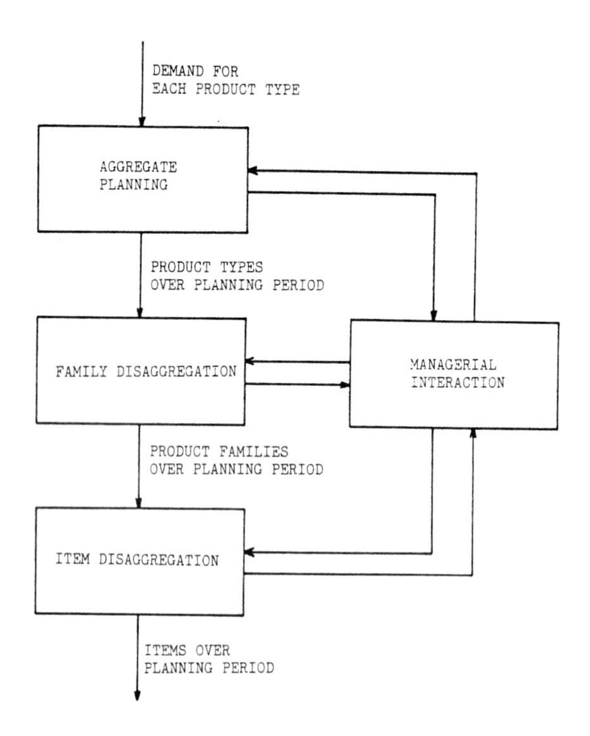

Fig. 2. Hierarchical planning system

expect an efficient methodology applied to FMSs.

Two recent survey papers, one by Buzacott and Yao (1982) and Sarin and Wilhelm (1983), are devoted entirely to the modeling of FMSs. It should be emphasized that most of the literature published on FMSs is based on queueing theory. An integer programming approach to the problems of FMSs has been underestimated. It seems that the integer programming approach embedded in an efficient information system can be used to model and efficiently solve many of the planning and scheduling problems, including the dynamic problems. Such an information system is discussed in Suri and Whitney (1984).

Based on the classical planning and scheduling approaches discussed in this paper, a methodology applicable to FMSs is outlined. The proposed methodology tries to incorporate optimization into the planning and scheduling decision process to a large extent. The problems being subject to optimization are of reduced sizes, either through aggregation or decomposition. Such an approach assures solution of these problems within a practically acceptable time. The new FMS planning and scheduling methodology is presented in Figure 3.

The aggregate planning (level 1 in Figure 3) is, to some degree, similar to the aggregate model of Hax and Meal (1982). To reduce the problem size, the products are grouped into product types. It is necessary, however, to modify Hax

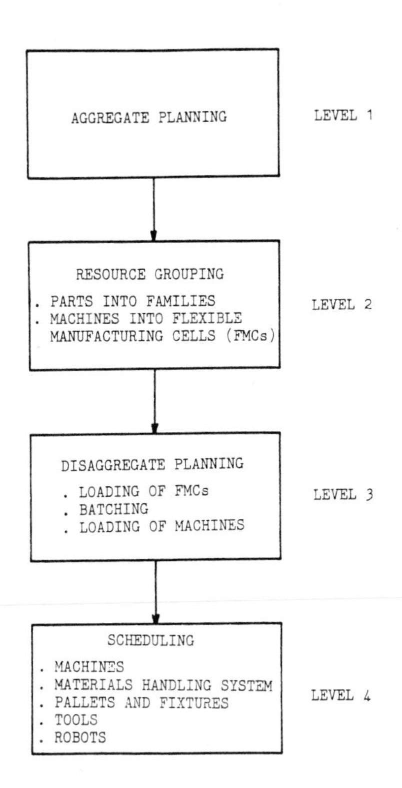

Figure 3. FMS planning and scheduling methodology

and Meal's (1975) model to make it applicable to FMSs. The modification includes both the objective function and the constraints. First of all, the workforce factor can be eliminted in the aggregate model of Hax and Meal (1975). A typical objective function in an FMS aggregate model would include the sum of inventory costs, production costs and costs representing production capacity. It should be stressed that the production capacity can be a decision variable in the aggregated FMS planning model.

In the classical systems the production capacity was varied typically by hiring and firing policies. In FMSs, it is possible to vary the production capacity only by changing the technological parameters; i.e. machining speed or feed rate. It should be remembered that this also influences the production costs.

Resource grouping (level 2 in Figure 3) is a very important issue in FMSs and is discussed mainly in the parts context in Kusiak (1984c). Grouping of parts and machines enables decomposition of the overall planning problems into subproblems. As indicated in Kusiak (1984c), grouping of machines into Flexible Manufacturing Cells (FMCs) is considered as a logical grouping opposite to the physical cellular concept in the classical manufacturing systems.

After the parts and the machines have been grouped, a new problem of allocating the part families to FMCs arises. This problem is solved in level 3 in Figure 3. In the classical Group Technology (GT) concept, parts have been grouped into families. Typically one part family would be assigned to one manufacturing cell. As mentioned previously, in an FMS process of grouping of machines into FMCs is of a dynamic nature. The modularity and flexibility of the materials handling system enables logical grouping of machines according to the current production needs.

Grouping of machines in FMSs becomes a software issue, rather than the hardware layout problem in the GT manufacturing concept. In the planning and scheduling methodology proposed in this paper, the part families concept is also different than in the classical GT systems. We should look at generation of part families as an agregation process. At this stage (level 2 in Figure 3) parts can be aggregated subject to similarity of

- tools
- fixtures
- pallets
- robot grippers
- machines.

A number of different models could be formulated, however, the exact considerations would be beyond the scope of this paper. It should be stressed that this methodology assumes a relatively large number of families. On average, there will be more than one part family loaded to one FMC in level 3 in Figure 3. The objective function of the FMCs loading may have many different forms. As an example, the minimization of production costs or travel costs could be considered. The model's constraints could be concerned with the number of part families loaded on an FMC, number of tools, number of pallets, etc.

In level 3, after the FMC loading problem has been solved, it may be necessary to split parts into batches. This, however, depends on the length of the planning period of level 1 and order sizes. The short planning period and the small size of orders may result in small batch sizes. If this would not occur, the batching model related to the aggregated model of Bitran et al.(1981) could be formulated for part families.

After the FMCs have been loaded with part families, another optimization problem arises within each FMC. There is always some flexibility in assigning operations to machines. The problem of loading of machines with operations (level 3 in Figure 3) can be modeled in a number of different ways. Loading problems of this nature have been formulated in Kimemia and Gerschwin (1980), Kusiak (1984b) and Stecke (1983).

One of the advantages of the proposed methodology is that the resulting scheduling problems (level 4 in Figure 3) can become relatively easy to solve. The loading performed in level 3 may decompose the multi-machine scheduling problem, for example into a one- or two-machine scheduling problem. The number of machines involved in the resulting scheduling problem depends on the loading problem formulation (see Kusiak 1984b). As stated in Kusiak (1984a), there are other machine scheduling issues which would be considered in FMSs; namely, scheduling of:

- materials handling system
- pallets and fixtures
- tools
- robots.

MACHINE LOADING PROBLEM

One of the most important problems in the presented FMS planning and scheduling methodology is that of machine loading (see level 3 in Figure 3). There is extensive literature on the loading problem. One of the most recent surveys on this subject is presented in Stecke and Solberg (1981).

The classical loading problem falls into a class of problems known in the operations research literature as the assignment problem (Ross and Soland 1980), the generalized transportation problem (Pogány 1978), the bin packing problem (Jonson 1974), the multiple knapsack problem (Hung and Fisk 1978), and the loading problem (Deane and Moodie 1972, Christofides, Mingozzi and Toth 1979).

Let us concentrate on two of the above mentioned formulations, namely:

- the generalized assignment problem
- the generalized transportation problem.

Given appropriate interpretation to these two operations research problems, they become loading problems in the manufacturing systems. In manufacturing systems, typically, the batches of parts are being loaded. Of course, there are a number of operations to be performed on each part. In this paper, for the conceptual and the notational convenience loading of batches of operations is considered.

Loading Model Based on the Generalized Assignment Problem

This loading model is based on a formulation of the generalized assignment problem. In order to present this model let the following notations be introduced:

I set of batches of operations to be processed

J set of stations

T_{ij} time of processing batch i on station j, for each $i \in I$ and $j \in J$

C_{ij} cost of processing batch i on station j, for each $i \in I$ and $j \in J$

b_j processing time available on station j, for each $j \in J$

$$y_{ij} = \begin{cases} 1 & \text{if batch i is processed on station j} \\ 0 & \text{otherwise} \end{cases}$$

The objective is to minimize the total sum of the processing costs,

$$\text{(MA)} \quad \min \sum_{i \in I} \sum_{j \in J} C_{ij} y_{ij} \tag{1}$$

$$\text{s.t.} \quad \sum_{j \in J} y_{ij} = 1 \quad , \text{ for each } i \in I \tag{2}$$

$$\sum_{i \in I} T_{ij} y_{ij} \leq b_j \quad , \text{ for each } j \in J \tag{3}$$

$$y_{ij} = 0,1 \quad , \text{ for each } i \in I \text{ and } j \in J \tag{4}$$

Constraints (2) ensure each batch of operations is processed on exactly one station. Constraints (3) ensure each station availability time cannot be exceeded.

One of the most effective algorithms for solving model MA has been developed by Ross, Soland and Zolteners (1980). Their algorithm is of the branch and bound type, with one of the bounds based on Lagrangian relaxation. They reported on computational results for many generalized assignment problems.

As an example consider the problem for $m = 30$ and $n = 40$ which was solved in 0.207 seconds, where m is number of stations and n is number of batches.

Loading Model Based on the Generalized Transportation Problem

Consider the assigning of a set I of batches of operations to a set J of stations. Each of these stations is capable of processing any batch of operations with different efficiency.

Let us denote:

- t_{ij} unit time of processing an operation from batch i on station j, for each $i \in I$ and $j \in J$
- c_{ij} unit cost of processing an operation from batch i on station j, for each $i \in I$ and $j \in J$
- a_i required number of operations from batch i, for each $i \in I$
- b_j processing time available on station j, for each $j \in J$
- x_{ij} number of operations from batch i to be processed on station j, for each $i \in I$ and $j \in J$

The loading model to MT can be formulated as follows:

$$\text{(MT)} \quad \min \sum_{i \in I} \sum_{j \in J} c_{ij} x_{ij} \tag{5}$$

$$\text{s.t.} \quad \sum_{j \in J} x_{ij} = a_i \quad , \text{ for each } i \in I \tag{6}$$

$$\sum_{i \in I} t_{ij} x_{ij} \leq b_j \quad , \text{ for each } j \in J \tag{7}$$

$$x_{ij} \geq 0 \; , \quad , \text{ for each } i \in I \text{ and } j \in J \tag{8}$$

Constraints (6) specify that the required number of operations from batch i is equal to a_i. Constraints (7) ensure that the time available on each station j is not exceeded. There has been only a few attempts to solve the model MT reported in the literature. Charnes and Cooper (1954) presented the stepping stone algorithm for solving model (MT), which has been generalized by Eisemann (1964).

Machine Loading Models Applicable to FMSs

Based on the previous considerations we will present four new machine loading models applicable to FMSs. In formulation of these models we will try to stick to the following two principles:

- formulate the models in a simple way
- use the practical assumptions.

Loading Model M1. One of the requirements in many of the FMSs is that each operation could be processed on more than one station. The loading model M1 presented below incorporated this requirement.

To formulate model M1, in addition to notation introduced in model MT, define the following parameters:

d_{ij} maximum allowable time for processing set of operations from batch i on station j, for all $i \in I$ and $j \in J$

n_i maximum number of stations the set of operations from batch i can be processed on, for each $i \in I$.

The formulation of model M1 is as follows:

$$\text{(M1)} \quad \min \sum_{i \in I} \sum_{j \in J} c_{ij} x_{ij} \qquad (9)$$

$$\text{s.t.} \quad \sum_{j \in J} x_{ij} = a_i \quad , \text{ for each } i \in I \qquad (10)$$

$$t_{ij} x_{ij} \le d_{ij} y_{ij} \quad , \text{ for each } i \in I \text{ and } j \in J \qquad (11)$$

$$\sum_{j \in J} y_{ij} \le n_i \quad , \text{ for each } i \in I \qquad (12)$$

$$x_{ij} \ge 0, \text{ integer} \quad , \text{ for each } i \in I \text{ and } j \in J \qquad (13)$$

$$y_{ij} = 0, 1 \quad , \text{ for each } i \in I \text{ and } j \in J \qquad (14)$$

Loading Model M2. There are constraints which are usually of no concern to management of the classical manufacturing systems, but which play an important role in FMSs. These constraints are concerned with a limited tool magazine capacity. In order to formulate a model which will incorporate a tool limit constraint, let us introduce the following parameters:

k_{ij} space occupied by the tool required for manufacturing of operations from batch i on station j, for each $i \in I$ and $j \in J$

f_j station j tool magazine capacity, for each $j \in J$.

Formulation of the loading model M2 is as follows:

$$\text{(M2)} \quad \min \sum_{i \in I} \sum_{j \in J} c_{ij} x_{ij} \qquad (15)$$

$$\text{s.t.} \quad \sum_{j \in J} x_{ij} = a_i \quad , \text{ for each } i \in I \qquad (16)$$

$$t_{ij}x_{ij} \leq d_{ij}y_{ij} \quad \text{, for each } i \in I \text{ and } j \in J \tag{17}$$

$$\sum_{i \in I} k_{ij}y_{ij} \leq f_j \quad \text{, for each } j \in J \tag{18}$$

$$\sum_{j \in J} y_{ij} \leq n_i \quad \text{, for each } i \in I \tag{19}$$

$$x_{ij} \geq 0, \text{ integer} \quad \text{, for each } i \in I \text{ and } j \in J \tag{20}$$

$$y_{ij} = 0, 1 \quad \text{, for each } i \in I \text{ and } j \in J \tag{21}$$

Loading Model M3. Apart from the constraints introduced in models M1 and M2, there are also constraints associated with the length of tool life. To formulate loading model M3, in addition to the notation in model M2, let us denote

r_{ij} expected length of life of a tool applied in processing operations from batch i on station j.

$$\text{(M3)} \quad \min \sum_{i \in I} \sum_{j \in J} c_{ij}x_{ij} \tag{22}$$

$$\text{s.t.} \quad \sum_{j \in J} x_{ij} = a_i \quad \text{, for each } i \in I \tag{23}$$

$$\sum_{i \in I} t_{ij}x_{ij} \leq b_j \quad \text{, for each } j \in J \tag{24}$$

$$t_{ij}x_{ij} \leq r_{ij}y_{ij} \quad \text{, for each } i \in I \text{ and } j \in J \tag{25}$$

$$\sum_{j \in J} y_{ij} \leq n_i \quad \text{, for each } i \in I \tag{26}$$

$$x_{ij} \geq 0, \text{ integer} \quad \text{, for each } i \in I \text{ and } j \in J \tag{27}$$

$$y_{ij} = 0, 1 \quad \text{, for each } i \in I \text{ and } j \in J \tag{28}$$

Loading Model M4. This model incorporated features of model M2 and M3

$$\text{(M4)} \quad \min \sum_{i \in I} \sum_{j \in J} c_{ij}x_{ij} \tag{29}$$

$$\text{s.t.} \quad \sum_{j \in J} x_{ij} = a_i \quad \text{, for each } i \in I \tag{30}$$

$$\sum_{i \in I} t_{ij}x_{ij} \leq b_j \quad \text{, for each } j \in J \tag{31}$$

$$\sum_{i \in I} k_{ij}y_{ij} \leq f_j \quad \text{, for each } j \in J \tag{32}$$

$$t_{ij}x_{ij} \leq r_{ij}y_{ij} \quad \text{, for each } i \in I \text{ and } j \in J \tag{33}$$

$$\sum_{j \in J} y_{ij} \leq n_i \quad \text{, for each } i \in I \tag{34}$$

$$x_{ij} \geq 0, \text{ integer} \quad \text{, for each } i \in I \text{ and } j \in J \tag{35}$$

$$y_{ij} = 0, 1 \quad \text{, for each } i \in I \text{ and } j \in J \tag{36}$$

SOLVING THE FMS LOADING MODELS

The presented FMS loading models (M1-M4) belong to a class of mixed integer (integer and boolean) programming problems. They can be solved by a general mixed integer programming techniques, i.e. cutting plane and branch and bound.

In practice, however, it may be difficult to obtain optimal solutions to the FMS loading models in an acceptable time, due to their large size (for example $m = 20$, $n = 2000$). Taking into account the FMS environment, in many cases it is satisfactory to generate a feasible solution. Computational experience with models related to M1-M4 indicates that such a feasible solution, in fact, may be very close to the optimal one. This is mainly due to the data structure associated with the existing FMSs; namely, the cost coefficients c_{ij} in the objective function are typically uniformly distributed.

These is also one more FMS feature which should be mentioned, namely the planning horizon corresponding to the loading models. The models discussed (M1-M4) are static FMS models, but they can be applied in dynamic situations as well. In the static applications (long planning horizon, for example 24 hours), the time to solve these models is not a crucial factor. One can imagine a situation where the long horizon results become invalid, because of FMS disturbances; i.e. machine failure. This generates a dynamic situation which requires recomputation of the loading model. In this case the length of the computing time becomes a very crucial factor. In such a case, one might be satisfied with the suboptimal solution.

Taking into account the features of the FMS environment discussed, a class of subgradient algorithms seems well suited for solving the four FMS loading models. These algorithms, typically generate a feasible solution in a modest computing time. It may take a considerably longer time to find an optimal solution; or, in many cases to confirm the optimality of the previously found feasible solution. A class of subgradient algorithms also has an advantage of indicating the quality of any feasible solution found. A relative distance of a feasible solution from the optimal one is very often used as a stopping criterion.

To illustrate the efficiency of a subgradient algorithm, consider the following formulation of the segregated storage problem.

$$\text{(PS)} \quad \min \sum_{i \in I} \sum_{j \in J} c_{ij} x_{ij} \tag{37}$$

$$\text{s.t.} \quad \sum_{j \in J} x_{ij} = a_i \quad , \text{ for all } i \in I \tag{38}$$

$$x_{ij} \leq b_j y_{ij} \quad , \text{ for all } i \in I \text{ and } j \in J \tag{39}$$

$$\sum_{i \in I} y_{ij} \leq 1 \quad , \text{ for all } j \in J \tag{40}$$

$$x_{ij} \geq 0, \text{ integer} \quad , \text{ for all } i \in I \text{ and } j \in J \tag{41}$$

$$y_{ij} = 0, 1 \quad , \text{ for all } i \in I \text{ and } j \in J \tag{42}$$

This problem is a special case of the loading model M1.

The computational results of solving the problem (PS) on the CDC CYBER 170/720 by the subgradient algorithm are demonstrated in Table 1.

TABLE 1

Problem Size		Average Number	Average Computing
m	n	of Iterations	Time in CPU sec.
20	10	20	0.50
25	20	9	0.84
25	25	15	1.29
30	20	4	0.88
40	30	16	2.9

The average number of iterations and the average computing time in Table 1 are reported for optimal solutions to the problem PS. A more detailed computational analysis is presented in Gunn and Kusiak (1983).

CONCLUSIONS

In the first part of this paper the planning and scheduling methodologies of the classical manufacturing systems were discussed. As these methodologies do not satisfy the needs of FMSs, a new planning and scheduling concept was outlined. This concept should fill the existing gap in the theory and applications of FMSs. It may also indicate the future research directions towards more efficient control strategies in FMSs.

The second part of this paper shows a relationship between FMS machine loading models and some of the known operations research models. Four new formulations of the FMS loading problem are presented. The strength of these formulations is in their linear structure. There are good prospects for applications of these models in the existing FMSs. This is mainly due to the constraints encompassing a large variety of industrial situations.

ACKNOWLEDGEMENT

This research was partially supported by the Natural Sciences and Engineering Research Council of Canada.

REFERENCES

1. G.R. Bitran, E.A. Hass and A.C. Hax, Hierarchical production planning system: A two-stage system, Operations Research, 30, (1982), 232-251.
2. G.R. Bitran, E.A. Hass and A.C. Hax, Hierarchical production planning: A single stage system, Operations Research, 29, (1981). 717-743.
3. G.R. Bitran and A.C. Hax, On the design of hierarchical production planning systems, Decision Sciences, 8, (1977), 28-55.
4. J.A. Buzacott and D.D. Yao, Flexible manufacturing systems: A review of models, Working Paper No. 82-007, Department of Industrial Engineering, University of Toronto, (1982).
5. A. Charnes and W.W. Cooper, The stepping stone method of explaining linear programming calculations in transportation problems, Man. Sci., (Oct. 1954) 49-69.

6. N. Christofides, A. Mingozzi and P. Toth, Loading problems, in Christofides, N. et al. (Eds.) Combinatorial Optimization, Wiley, New York (1979) 425 pp.
7. R.H. Deane and C.L. Moodie, A dispatching methodology for balancing workload assignments in job shop production facility, AIIE Transactions, 4, (1972) 277-283.
8. M.A.H. Dempster, A stochastic approach to hierarchical planning and scheduling, in Dempster, M.A.H. et al. (Eds.) Deterministic and Stochastic Scheduling, D. Reidel Publishing Company, Dordrecht, Holland, 1982, 419 pp.
9. M.A.H. Dempster, M.L. Fisher, B. Legewang, L. Janssen, J.K. Lenstra and A.G.H. Rinnooy Kan, Analytical evaluation of hierarchical planning systems, Operations Research, 29, (1981), 707-717.
10. K. Eisemann, The generalized stepping stone method for the machine loading model, Man. Sci., 11, (1964), 154-176.
11. S.E. Elmaghraby, The economic lot scheduling problem (ELSP): Review and extensions, Management Science, 24, (1978), 587-598.
12. L.F. Gelders and L.N. VanWassenhove, Production planning: A review, European Journal of Operational Research, 7, (1981), 101-110.
13. E.A. Gunn and A. Kusiak, A methodology for application of Fenchel's duality theory to large scale problems. Working Paper No. 05/82, Dept. of Industrial Engineering, Technical Unversity of Nova Scotia, Canada, (1982).
14. S. Graves, A review of production scheduling, Operations Research, 29, (1981) 646-675.
15. A.C. Hax and H.C. Meal, Hierarchical integration of production planning and scheduling, in Geisler, M.A. Ed., TIMS Studies in Management Sciences, 1, Logistics, North-Holland, Amsterdam, (1975), 53-69.
16. M.S. Hung and J.C. Fisk, An algorithm for 0-1 multiple knapsack problems, Naval Research Log. Quarterly, 25 (1978), 571-579.
17. J.G. Kimemia and S.B. Gerschwin, Multicommodity network flow optimization in flexible manufacturing systems, Report No. ESL-FR-834-2, Electronic Systems Laboratory, MIT, Cambridge, (1980).
18. A. Kusiak, Analysis of flexible manufacturing systems, Working Paper #09/83, Department of Industrial Engineering, Technical University of Nova Scotia, Halifax, Nova Scotia, (1983).
19. A. Kusiak, Flexible manufacturing systems: A structural approach, Working Paper No. 4/84, Department of Industrial Engineering, Technical University of Nova Scotia, Halifax, Nova Scotia, (1984a)
20. A. Kusiak, Loading models in flexible manufacturing systems, Proceedings of the 7th International Conference on Production Research, Windsor, Ontario, (1984b), 641-647.
21. A. Kusiak, The part families problem in flexible manufacturing systems, Working Paper No. 06/84, Department of Industrial Engineering, Technical University of Nova Scotia, Halifax, Nova Scotia, (1984c).
22. D. Jonson, Fast algorithm for bin packing, Journal of Computer and System Science, 8, (1974), 272-314.
23. J.K. Lenstra, Sequencing by enumerative methods, Mathematical Centre Tract, Matematisch Centrum at Amsterdam, (1977).
24. Z. Pogány, An algorithm for solving the generalized transportation problem, Proceedings of the 8th IFIP Conference on Optimization Techniques, Würzburg, September 5-9, (1977), Springer Verlag, New York, (1978).
25. G.T. Ross and R.M. Soland, A branch and bound algorithm for the generalized assignment problem, Math. Programming, 18, (1975) 91-103.
26. G.T. Ross, R.M. Soland and A.A. Zolteners, The bounded interval generalized assignment model, Nav. Res. Log. Quart., 27, (1980) 625-633.
27. K.E. Stecke and J.J. Solberg, The Optimal Planning of Computerized Systems. The CMS Loading Problem, Report No. 20, School of Industrial Engineering, Purdue University, (1981).
28. S.C. Sarin and W.E. Wilhelm, Models for the design of flexible manufacturing systems, Proceedings of the Annual IIE Conference, Louisville, KY., (1983), 564-574.

29. E.A. Silver, Operations research in inventory management: A review and critique, Operations Research, 29, (1981), 628-645.
30. K.E. Stecke, Formulation and solution of nonlinear integer production planning problems for flexible manufacturing systems, Management Science, 29, (1983), 273-288.
31. R. Suri and C.K. Whitney, Decision support requirements in flexible manufacturing, Journal of Manufacturing Systems, 3, (1984).
32. L.N. VanWassenhove and M.A. DeBodt, Capacitated lot sizing for injection molding: A case study, Working Paper #80-26, Katholieke Universiteit Leuven, (1980).

THE ADAPTIVE CONTROL OF MACHINING IN FLEXIBLE MANUFACTURING SYSTEMS

PETER F. McGOLDRICK

Department of Production Engineering and Production Management
University of Nottingham, Nottingham NG7 2RD, (United Kingdom)

ABSTRACT

Flexible Manufacturing Systems achieve high levels of productivity through reliance on several interlinked technological features; high production rate machine tools and processes are linked by automatic, and usually robotic handling, with the whole being under computer supervision and control.

However flexible and intelligent such systems are, or indeed might become, their overall productivity is still largely controlled by the throughput ability of the particular manufacturing process or processes at the heart of the system. Most such systems are, and are likely to remain, centred around machining processes.

Intelligence in machining has failed to keep pace with that of the rest of the manufacturing system and it is argued that much more research and development effort is needed if this situation is to be rectified; true self-optimising adaptive controllers are now needed if the real productivity potential of Flexible Manufacturing Systems is to be realised.

INTRODUCTION

Adaptive control is considered to exist under two headings, adaptive control constraint (ACC) and adaptive control optimisation (ACO), and is categorised according as to whether technological or geometrical factors are being used for control.

ACC technological systems are now relatively common and are available as original or retro-fit equipment on a variety of metal-cutting machine tools; typical systems are described. ACO technological systems have been the subject of much fundamental research but the non-availability of appropriate transducers is a major stumbling block to future commercial exploitation and industrial use.

From a fundamental production viewpoint, it would be much more satisfactory if adaptively controlled manufacturing systems relied on geometrical factors as the basis for their operation since it is features such as size and surface finish which control the ultimate output of the system - the product.

ADAPTIVE CONTROL

Adaptive Control (AC) has no unique meaning in scientific parlance but a more rigid differentiation between two sorts of system has helped the situation from the point of view of the manufacturing engineer. It has been defined (ref. 1) as:

> *"a control system that measures certain output process variables and uses these to control speed and/or feed. Some of the process variables that have been used in adaptive control machining systems include spindle deflection or force, torque, cutting temperature, vibration amplitude, and horsepower. In other words, nearly all the metal-cutting variables that can be measured have been tried in experimental adaptive control systems."*

This is in fact a very narrow definition since it precludes consideration of control of the geometry of the product - a factor which will be discussed later.

The subdivision into ACC and ACO systems is a useful one to consider in some detail.

Adaptive Control Constraint (ACC)

These systems are available on many commercial controllers; being produced as either standard or optional original equipment or as a piece of retro-fitted hardware.

In these systems some parameter of the process is constrained to a particular value which is input to the simple feedback control system which is regulating that process. Examples of systems of this sort are; the measurement of cutter spindle bending which is maintained at a ceiling value; the measurement of the current supplied to a direct current motor - in this way the motor torque is known and can be constrained to a pre-set value.

It can quite legitimately be argued that in the mathematical sense of the word these are not adaptive control systems at all - they are simple feedback controllers. The constraint level selected must by the nature of the present state of knowledge of the technology of the process be little more than an educated guess; the level could be such that, if it were exceeded, then damage to the machine or workpiece would result but such extremes of conditions are not likely to be attractive in the commercial sense.

Adaptive Control Optimisation (ACO)

These systems, which currently only exist as equipment in research laboratories, evaluate some Index of Performance which is then optimised. This Index typically relates metal removal rate to tool wear rate and thus, as will be discussed later, relies on the ability to measure, or otherwise monitor, tool wear in-process.

Since the primary objective of any machining or other manufacturing process is to output a product, then it is surely entirely logical to argue that metal removal is merely a means of achieving that objective; the tenet that most ACO system designers appear to hold is that it is an objective in itself. A very misleading priority for system design could be the result of such muddled thinking.

TECHNOLOGICAL ADAPTIVE CONTROL

The problem with systems designed under this sub-group has already been mentioned since here one chooses to control the parameters of the process and not the configuration of the product. Despite the weight of brainpower and the vast research effort applied, our understanding of the simplest of metal cutting operations has advanced very little in forty years. One major U.K. tool manufacturer has said (ref. 2) that they are now in a position to predict the mechanism by which a particular tool will fail when machining a particular workpiece - but not when that failure will occur. Even the establishment of machining data banks (ref. 3) is unlikely to prove the answer since it is universally recognised that the scatter of results observed in replications of what are nominally the same experimental conditions means that one is dealing with a stochastic and not a deterministic process. It is the case that if tool wear were absolutely predictable then on-line control would not be needed; the machine tool could be pre-programmed to a known behaviour pattern.

Many of these problems would of course be alleviated if it were possible to measure tool wear in-process. Again much work has been devoted to this area (ref. 4) but progress has been slight - and very slow.

If one moves away from direct measurement of tool wear into the area of measuring its influence on other parameters of the process - force, torque, power, bending and so on - then, for the same reasons, the picture is just as muddy and research effort just as unrewarding.

GEOMETRIC ADAPTIVE CONTROL

These systems have much appeal for production engineers in that they seek to control the thing that matters - the product; transducers are again a problem.

There are commercial plunge cylindrical grinding machines available which sense the size of the ground surface using a callipers arrangement and use this data to control the wheel in-feed rate; with this exception, no machines are known to exist.

The problems of developing sensors which can detect size, surface finish and other geometric features at machining speeds and under machining conditions are enormous; however some, albeit limited, progress has been made (ref. 5, 6).

RESEARCH AT THE UNIVERSITY OF NOTTINGHAM

This has concentrated around milling in general and end-milling in particular (ref. 7, 8, 9, 10).

The process is of considerable interest in its own right but is also of major importance since it is a very common process throughout the whole of manufacturing industry. Because of the complexity of the tooling used and the geometry of the cut produced, it is as complex as almost any metal-cutting process currently in use; resolution of the problems of the adaptive control of milling will make it easier to solve the same problem for processes such as turning and grinding where the geometry is much more straight-forward.

The extensive use of milling-type, numerically controlled machining centres in unmanned production facilities has increased the impetus to solve the problems of the adaptive control of this process.

MACHINE TOOL

An old ROSSI manual vertical milling machine was modified for the adaptive control research (ref. 7) by the fitting of two thyristor (Silicon Controlled Rectifier - SCR) drives with their associated direct current motors.

One of these motors was fitted to the spindle drive, actually driving via a pair of toothed pulleys and a rubber timing belt. This enables various ranges of cutting speeds to be investigated by the simple expedient of changing the pulleys used. In the configuration usually used, the system has constant torque up to a spindle speed of about 300 revolutions per minute and then a constant power (of about 1.3 kiloWatt) up to about 1000 revolutions per minute; speed is naturally continuously variable in the range 0 to 1000 revolutions per minute.

The table traverse was also modified to accept the second drive directly coupled to a recirculating ballscrew. The table feedrate was set to be continuously variable in the range 0 to 500 millimetres per minute.

CONTROLLER

A DEC PDP 11/15 mini-computer with associated analogue to digital conversion circuit boards which had been in use for other work in the department was pressed into service. The machine's power and sophistication were not needed but they did enable fast processing using a high-level language to be undertaken.

Since suitable analogue to digital routines had already been written in BASIC and these had proved reliable and easy to use in connection with other work (ref. 11) they were again called up as sub-routines within the optimising routines discussed in this paper. All writing of programs was thus undertaken in BASIC.

[Recently a DEC PDP 11/34 has replaced the elderly 11/15 and work is continuing with this almost direct but more powerful substitute controller.]

OPERATIONAL PRINCIPLES

The electronic circuit boards driving both the S.C.R. drives were interrogated to establish the existing value of both the motor current and speed in the case of the spindle and speed in the case of table traverse; thus signals representing spindle motor torque and speed (and hence power) and table feedrate were established and fed downline to the computer. This information is converted into digital form which is then used as input into the optimisation routines. The output from the optimisation is converted from digital to analogue form and then instructs the motor drive circuits thus closing the adaptive loop.

OPTIMISATION CRITERIA

These are discussed more fully elsewhere (ref. 12) but essentially the Index of Performance can be regarded as a third axis in a cartesian set where the other two axes represent the independent variables cutting speed and table feedrate. If a single value of cutting speed coupled with a value of feedrate gave a unique and time independent value of the Index of Performance then on-line control would not be needed - the conditions could be pre-set off-line. Because of the non-linear behaviour of tool wear - and, in the long term, its deterministically unpredictable behaviour - it would ultimately be necessary to have an on-line evaluation of the Index of Performance. The strategy of on-line measurement and evaluation was thus adopted from the outset even though it was in many of the early situations an unnecessary luxury.

A time variable function describing tool wear had been established using a weight loss method (ref. 4) and this was combined with an economic analysis governed by metal removal rate which enabled the adaptive loop to be closed.

FLEXIBLE MANUFACTURING SYSTEMS

The impetus to seek a solution to the problems faced by those engineers concerned with Adaptive Control could well come from the interest being shown across the world in Flexible Manufacturing Systems. These systems, consisting of a variety of machine tools - principally, but not exclusively, metal cutting machines - linked by a handling system and with the entire system under the control and supervision of a computer or hierarchy of computers, have the potential to revolutionise the organisation of manufacturing industry. Whilst it is a technology which is as much concerned with software as it is with hardware, it is the latter problems with which this paper is concerned.

Heavy system utilisation must be of paramount importance when one is considering technology which involves such high levels of capital investment but running alongside we have seen more and more over recent years, moves towards severe reductions in manning levels. Whether such moves are desirable from a social and indeed national economic viewpoint is a crucially important point, but one must presume from the number of such systems cropping up across the world, that management is judging that they are the way to proceed at company level.

REQUIREMENTS OF MINIMUM MANNED SYSTEMS

It is probably sensible to view this problem as that of looking at systems with levels of manning which are a minimum; it is extremely unlikely that, no matter how intelligent computers may get, we will ever see a totally unmanned manufacturing enterprise.

Machine tools must be protected against overload at all times, and this factor has long been recognised. The problem is however not one of protecting electrical systems against current surges and the like but rather of mechanical overload protection.

Tool breakage detection is clearly a first priority and much progress has been made but, rather as with tool wear systems, the stochastic nature of the problem has not been fully recognised. It is for example perfectly possible for a spiral twist drill to break with no change in torque, vibration level, noise level or tool signature. Given this, the problem of detecting tool failure, when there is no obvious physical parameter with which to assess that failure with total reliability, is enormous.

Equally those control systems which remember the time for which a particular cutting tool has been used and then utilise "sister-tool replacement" - replacement by an identical tool - must either be risky or conservative. Everybody involved in manufacture can cite a case where a tool which usually lasts hundreds of minutes fails in the first few; equally a system which assumes a particular life for a cutting tool must underestimate that life if an acceptable level of risk is to be attained. If one could assume that tool life is normally distributed, and I know of no evidence as to whether it is or it is not, then a system timed to operate at the mean will be conservative for about half of the time and coping with a failed tool the other half.

Torque and power overload is rather easier to cope with, since transducers are widely available, but again the problem of detecting a meaningful spike in a reading which fluctuates wildly with time should not be underestimated.

FUTURE DEVELOPMENTS IN ADAPTIVE CONTROL

The 'traditional' role of adaptive control - the economic evaluation and control of metal removal rate by the optimisation of speeds and feeds - seems likely to become secondary in importance to achieving high utilisation, which itself implies high reliability, in a set of systems which is going to feature less and less human involvement.

If one accepts that our understanding of the process in question is unlikely ever to be sufficiently good or reliable as to ensure that we can, with sufficient (however that might be defined) certainty, rely on the measure of some physical parameter or combination of parameters in order to control the process then there is only one way forward; control by evaluation of the product.

As mentioned earlier, evaluation of the metrological features of the product, on-line and at machining speeds is a problem to which there is, at present, no solution.

CONCLUDING REMARKS

Adaptive Control - particularly Geometric ACO systems - will have a place in flexible manufacturing systems of the future but research effort needs to be applied in increasing quantity.

ACKNOWLEDGEMENTS

The pioneering work of Munif Hijazi is gratefully acknowledged as is the support given to him by United Industries Corporation of Amman, Jordan; without his hard work and their financial assistance much less would have been achieved.

The Analogue to Digital routines called up by the adaptive controller were written by Dr. Vic Middleton of the Department of Mechanical Engineering, the University of Nottingham, to whom very grateful thanks are due.

REFERENCES

1 M.P. Groover, 'Automation, Production Systems, and Computer-Aided Manufacturing'. Prentice-Hall, New Jersey, United States of America, 1980.
2 Private communication.
3 PERA Research Report 302. 'Machining Data Bank'. Production Engineering Research Association, Melton Mowbray, United Kingdom, August 1976.
4 P.F. McGoldrick and M.A.M. Hijazi. 'The Use of a Weighing Method to Determine a Tool Wear Algorithm for End-Milling'. Proc.20th. Int.M.T.D.R. Conf., Birmingham 1979; Macmillan Press, London, United Kingdom, 1980.
5 R. Carter. 'On-Line Measurement of Milling Surface Finish'. B.Sc. Thesis, University of Nottingham, May 1980.
6 H.E.A. Hussein. 'Size Measurement in Turning'. B.Sc. Thesis, University of Nottingham, May 1980.
7 M.A.M. Hijazi. 'The Adaptive Control of End-Milling'. Ph.D. Thesis, University of Nottingham, September 1978.
8 S.J. Bessaibes. 'The On-Line Measurement of Cutting Forces in End-Milling'. B.Sc. Thesis, University of Nottingham, May 1979.
9 A. Geday. 'On-Line Optimisation of an Adaptively Controlled Milling Process'. B.Sc. Thesis, University of Nottingham, May 1979.
10 C.E. Englezos. 'On-Line Optimisation of an Adaptively Controlled Milling Process'. B.Sc. Thesis, University of Nottingham, May 1981.
11 T.C. Goodhead, P.F. McGoldrick and J.J. Crabtree. 'Automatic Detection of and Compensation for Alignment Errors in Machine Tool Slideways'. Proc.18th Int. M.T.D.R. Conf., London 1977; Macmillan Press, London, United Kingdom, 1978.

12 P.F. McGoldrick, M.A.M. Hijazi and A. Geday. 'Optimisation Strategies for Adaptively Controlled Milling'. Proc. of Prolamat 82 (5th International Conference on Programming Research and Operations Logistics in Advanced Manufacturing Technology), Leningrad, May 1982; published as 'Advances in CAD/CAM', editors T.M.R. Ellis and O.I. Semenkov, North-Holland Publishing Company, Amsterdam, The Netherlands, 1983.

IDEA AND PRACTICE OF FLEXIBLE MANUFACTURING SYSTEM OF TOYOTA

ATUSHI MASUYAMA
TOYOTA MOTOR CORPORATION, JAPAN

ABSTRACT

A definition of Flexible Manufacturing System has not been necessarily clarified yet; An understanding, and an objective of its application are different in a variety of industries. In Toyota, on the basis of market-oriented production, FMS is understood as follows.

(1) The ultimate objective of applying FMS is to provide a manufacturing system which may flexibly respond to changes in a market. The flexible response means to supply timely a product as demanded by a customer.

(2) Therefore, we should understand FMS as an extensive system from a product design through a product distribution, but not as a system limited to a process in manufacturing.

In order to get all subsystems to work out as an organic whole, we believe that a management system should be prepared concurrently.

A general idea of FMS (FMMS), which is conducted in Toyota, is illustrated from such viewpoints as need, scope, objective, and measurement of evaluation. We also introduce various activities relating shortening the lead time in production, and in production planning and production ordering.

1. INTRODUCTION

Economic activities have long been slow on a worldwide level. It is the most critical point for manufacturers to seek, without a half and half attitude a system to produce at the least cost only as much quantity as surely to be sold so that these manufactures may win the tough business race and survive under such circumstances. It seems essential to perceive accurately the condition of the market and to supply what is demanded by the market, with a short lead time and at a low cost.

Recently, FMS has been in the spotlight. The number of companies that have introduced FMS, or have been examining doing so is increasing, which proves that many companies realize importance and difficulty of flexibly responding to changes in the market. The perception of "flexible response", however, seems to vary among companies.

In this paper, we will introduce our insight into FMS and our practical activities.

2. WHY IS "FLEXIBILITY" NECESSARY?

— Environment Surrounding Auto Industry —

2.1 Market trend

Observing the demand of Japanese market trend of the last twenty years, there is a drastic difference between the first decade of that period and the second. The turning point was the oil crisis in 1973: the first decade can be expressed as producer oriented days in which products are usually sold out, meanwhile the second decade as consumer oriented days in which producers must be aware of over-production. (Fig. 1)

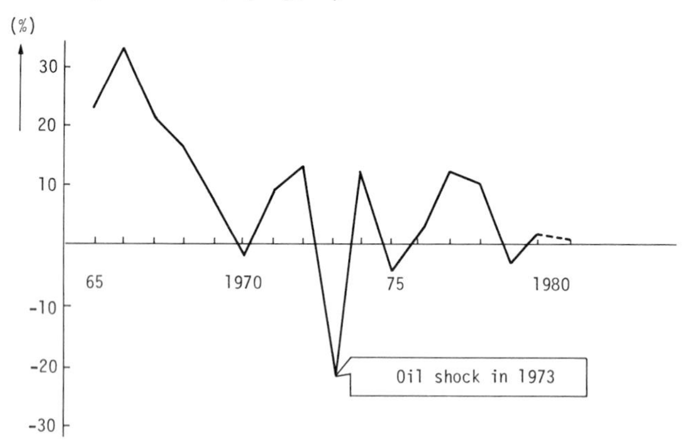

Fig. 1 Annual Growth Rate - Vehicle Market in Japan

On the other hand, the outlook of export business is not bright either on account of the export restraint, stemmed from the trade imbalance.

Since domestic and overseas markets become such conditions, the followings are essential to holding a dominant position in the market:

◦ Thoughtful expansion of product's specification so as to satisfy any customer's choice.

◦ Rapid response to a fluctuation in the quantity of the gross demand, and/or that in the quantity of various models.

◦ To conduct a smooth model change over.

In other words, it is required to respond to the market change with a short notice as well as to provide numerous end items. (Table 1)

	Variants	Quantity	Quantity/Variants
Car Line A	3,700	63,000	17
B	16,400	204,000	12
C	4,500	53,000	12
D	7,500	44,000	6
Total/Average	32,100	364,000	11

Table 1 Number of Variants and Quantity, March-May, 1982

2.2 Introduction of new technology

New mechanisms of a car, new materials and new production engineering may be introduced for the sake of improvement of safety standard,fuel consumption rate, resistance against corrosion, and producibility. Such introduction requires modification in production systems and processes. It is essential that manufacturers have flexibility of accepting such modification quickly and efficiently.

3. IDEA OF FMS IN TOYOTA

3.1 Objective

As mentioned above, environment (the market and technology trend) surrounding automotive manufacturing seems to become more liable to drastic changes. Hence, we believe it important that not only manufacturing cells or plants are flexible but also the entire production system should be flexibly (promptly and economically) responsive to such changes. That is, we must invent and practice a system to organize and control manufacturing cells in which automated equipments (FMS in a narrow sense) are incorporated, which we call as Flexible Manufacturing and Management System (FMS in a broad sense). We are to make efforts to construct a flexible production system, which is based on the perception that customers decide the worth of commodities and the criterion is subject to change, depending upon circumstances.

3.2 The scope of subjects

In order to make the thorough manufacturing activities flexible, we should not consider only part of the whole process, but it is important to connect organically all relating manufacturing functions from product development planning through distribution of finished goods. (Fig. 2)

Focusing on production system, the following stages must carry out the functions to pursue flexibility. (Fig. 3)

- Production Planning
- Parts and Material Planning
- Fabrication and Assembly

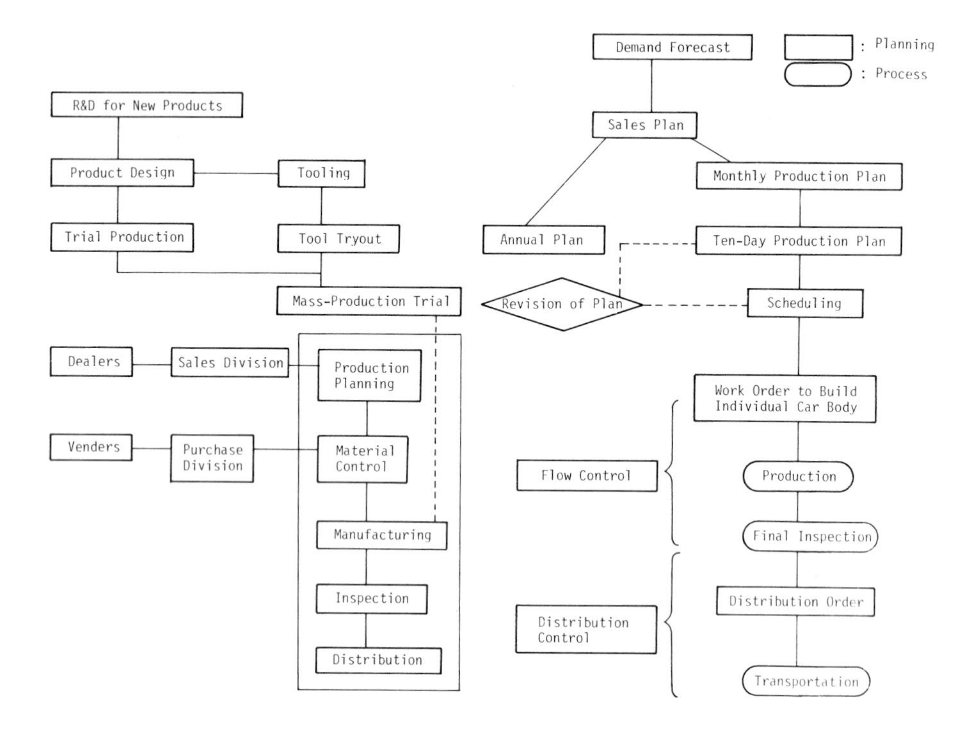

Fig. 2 Structure of the Functions Relating Production

Fig. 3 Production System

For instance, speaking about fabrication and assembly stage, making only machining shops flexible is hardly effective to improve the flexibility of the whole fabrication and assembly stage. It goes without saying that making only a machine or a machine-cell flexible is meaningless for the auto industry which consists of numerous processes to build a car. (Fig. 4)

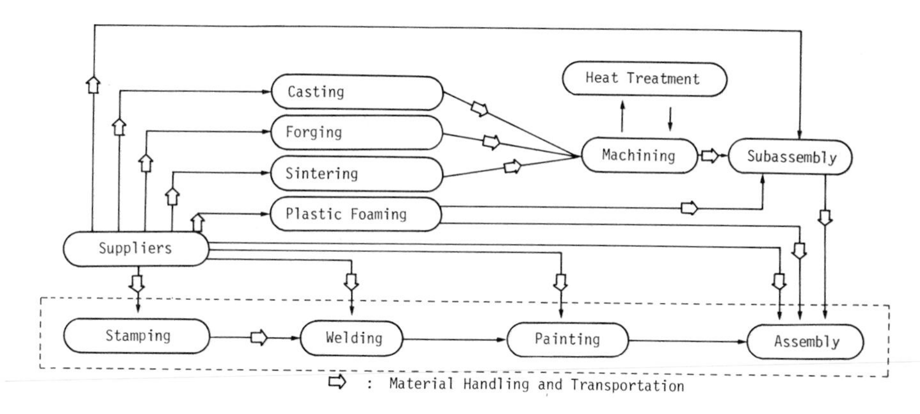

Fig. 4 Outline of Production Process

When the above "flexibility concept" is extended to vendors, the ideal FMS becomes attainable. We hereafter will confine our comments to flexibility of the production system.

3.3 Measure of evaluating flexibility

We evaluate the flexibility of production system, which is based on the following view points.

3.3.1 Prompt response to a change. We appraise that to what extent the leadtime from customer's order receipt though completion of products is minimized. The lead time can be classified as follows:

A. LT_i: The lead time of processing information and communication, including demand forecast, planning production, material requirement planning, production ordering etc.

B. LT_p: The lead time of manufacturing, including machining, assembly, inspection etc.

C. LT_t: The lead time of transpotation, including material handling and distributing finished goods.

To shorten each lead time is quite significant for flexible response to changes.

$$\circ\ LT \quad \left\{ \begin{array}{l} LT_i \\ LT_p \\ LT_t \end{array} \right. \Rightarrow \text{Min.}$$

The relation of the lead time and the accuracy of demand forecast, on which production planning is based, is generally as follows:

$$\hat{D}_{t,\ t-LT} \neq \hat{D}_{t,\ t-LT-1} \neq \cdots\cdots \neq \hat{D}_{t,\ t-1},$$
$$|\hat{D}_{t,\ t-LT} - D_t| > |\hat{D}_{t,\ t-LT-1} - D_t| \cdots\cdots > |\hat{D}_{t,\ t-1} - D_t|,$$

where $D_{t,\ t-LT}$ = demand in the period t, forecasted in the period t-LT,
D_t, = real demand in the period t.

In other words, forecast error can be minimzed when one makes a demand forecast for the period t in the period as close to the period t as possible, because a trend of the market can be best incorporated to the forecast value.

3.3.2 Economical response to a change. We appraise a production system, based upon how economically it can respond to changes. "Economical reponse" can be defined as a condition where minimized are inventory level, and the difference between the process capacity and the actual work load level as well.

- Inventory Level
- |Process Capacity - Work Load Level|

⇨ Min.

What we have to keep in mind is not to let the work load level approach to the process capacity, but we should consider the work load level is determined according to the requirements of the market.

4. ACTUAL ACTIVITIES

We would like to introduce activities which we have actually developed in the areas of production planning, production ordering and production so as to attain economically shortening the lead time. Each lead time of LT_i, LT_t consists of the following three factors:

- Processing Time (T_p)
- Setting up Time (T_s)
- Waiting Time (T_w)

} LT

Activities shortening the lead times are, in short, those activities to minimize T_p and T_s, and to eliminate T_w as well.

4.1 Shortening the lead time in manufacturing process

Toyota Production System began being developed in the 1950's. Since then, we have considered that the best way to shorten the lead time is to practice Toyota Production System without a half and half atitude. At Toyota, the lead time from the start of body assembly to the end of the final assembly line is about one day. Practicing Just-in-time and Autoactivation in each process, which are the two basic concept of Toyota Production System, enables us to satisfy customers' requirements and to minimize in-process inventory as well.

Just-in-time is, in short, for the purpose to complete the necessary products at the necessary time through synchronization of all processes, to process only the necessary components with short lead time. To practice Just-in-time is not difficult in the case that there is only a single specification in a product. However, a manufacturer has become required by the market to provide a product with various specifications or configurations. Demand of each specification or configuration is not stable, therefore manufactureres must have mixed-model lines dealing with demand fluctuation of such a product.

It is very difficult to manage Just-in-time in a mixed-model line. The following enable us to manage Just-in-time:

- Withdrawal by subsequent processes
- Manufacturing in a small lot
- Smoothed production

4.1.1 Withdrawal by subsequent processes. In Fig. 5, illustrated is that how production orders are given to each process in Toyota.

(i) Only the first station of body assembly line receives a build-information one by one from the production control room.

(ii) Other processes like sub-lines or distant stations receive relevant information as process of the main line proceeds, which enables the whole production system to synchronize even if disorder of equipment or quality problem requires stopping the assembly line or re-scheduling of the production sequence. The final assembly line withdraws sub-assemblies from where those are produced as much as the final assembly line has consumed to assemble cars. The preceding shop, where those subassemblies are produced, again gives the first station of the shop replenishment order as much as withdrawn by the subsequent shop, that is, the final assembly line.

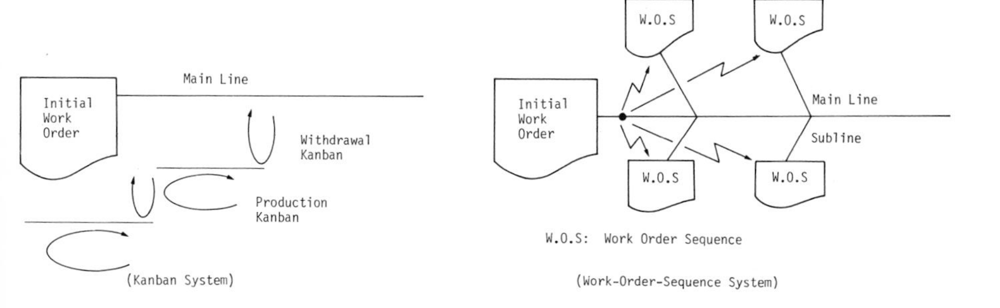

Fig. 5 Information and Material Flow in Production

This chain-like manner connects a series of processes in a multi-stage production system. Any process other than the final assembly line never produces sub-assemblies on a forecast basis. (Fig. 6) Kanban is used as a tool to materialize such withdrawal by a subsequent shop.

Withdrawal by a subsequent shop is called as "Pull System" which has been studied in comparison with "Push System" which is a more-general production control system. (Fig. 6)

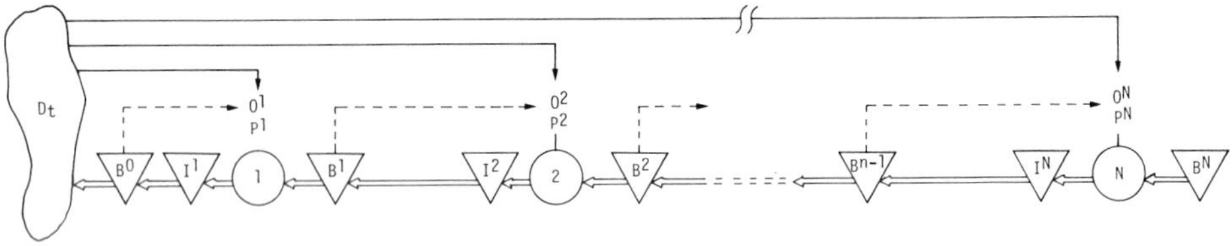

For all n, $O^n_{t:t+L^n+1}$

$$= \hat{D}_{t:t+LT^n+\ell} + \left(\sum_{\ell=1}^{L^n} \hat{D}_{t:t+LT^{n-1}+\ell} - \sum_{\ell=1}^{L^n} P_{t-\ell:t-\ell+L^n+1}\right) - B_t^{n-1} + S^{n-}$$

[Push System]

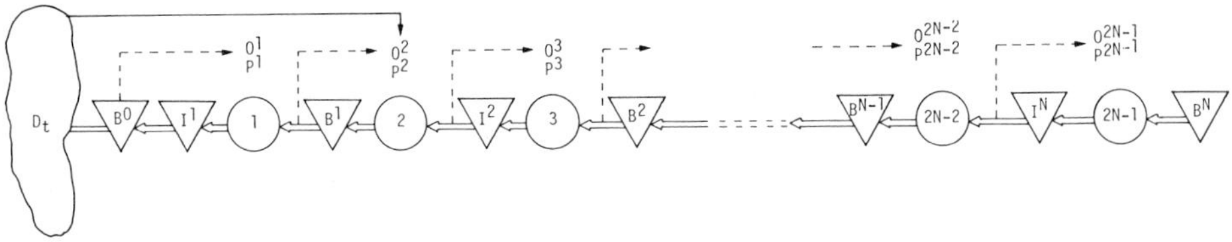

For n=1, $O^1_{t:t+L^1+1}$

$$= \hat{D}_{t:t+LT^1+1} + \left(\sum_{\ell=1}^{L^n} \hat{D}_{t:t+\ell} - \sum_{\ell=1}^{L^n} P_{t-\ell:t-\ell+L^1+1}\right) - B_t^{0} + S^{0}$$

For n > 1, $O_{t:t+L^n+1}$

$$= P^{n-1}_{t-L_2^{n-1}-L_1^n:t-L_1^n} + O^n_{t-1:t+L^n} - P^{n}_{t-L_2^{n-1}-L_1^n-1:t-L_1^n-1}$$

[Pull System]

Fig. 6 Push System and Pull System

The characteristic of "Pull System" is to produce only the necessary material required by subsequent shops. A simulation with mathematical models has proved that fluctuations in the subsequent shops is not amplified in the preceding shops. (Fig. 7)

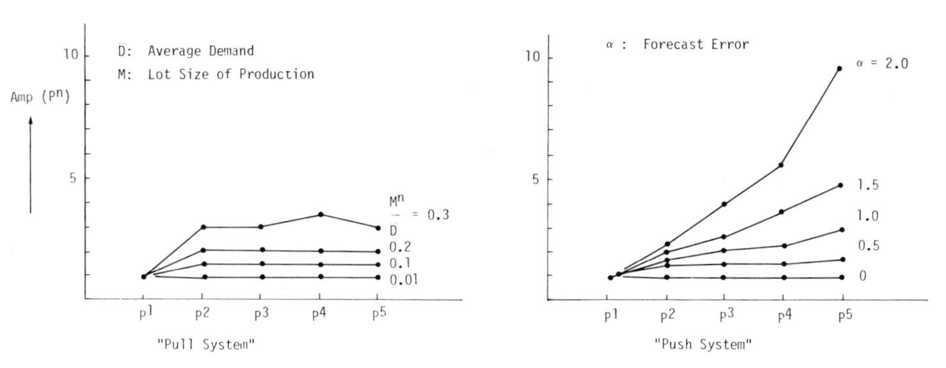

Fig. 7 Comparison between "Pull System" and "Push System"

4.1.2 Production with a small lot size. It is essential to minimize a lot size in production at each process in order to minimize the lead time and in-process inventory as well. Reduction of setup time is crucial to minimization of a lot size. The relation of total elapsed time to setup times, the number of setups, and lot sizes is expressed as follows:

T.E.T. = Total elapsed time

N_i = the number of setup times of item i

ts_i = setup time of item i

R_i = requirement of item i

tm_i = unit manufacturing time of item i

q_i = lot size of item i

$$N_i = R_i/q_i$$

$$\text{T.E.T.} = \Sigma N_i \cdot ts_i + \Sigma R_i \cdot tm_i$$

For a given set of R_i and T.E.T., it is necessary to reduce ts_i in order to reduce q_i. Once reduction of ts_i has been achieved, more items can be manufactured for a given T.E.T.

4.1.3 Smoothed production. When all procedures are linked by subsequent shop's withdrawals and small lot production, smoothing of product in the final assembly line with many variant's mix is essential to minimization of capacity requirement and to elimination of excessive inventory. Consequently, gross production quantity as well as consumption rate of individual material on the final assembly line must be smoothed in terms of daily output level as well as production sequence, computing the cycle time of variants.

4.1.4 Autoactivation. Autoactivation is defined as a mechanism in which equipment or operation is designed so as to stop when abnormal conditions occur, that is, we incorporate such a device to an equipment as would sense an abnormality and stop by itself, or we give all workers a power to stop the assembly line when they find any abnormality. What we call abnormalities here are defects, delay of operation, over production, machine trouble etc. The basis of Autoactivation concept are:

- To prevent over production
- Not to release defects to the subsequent shops
- To visualize abnormality so as to take prompt measures and to have all workers join improvement activities, which greatly contributes toward shortening the lead time in production.

4.2 Shortening the lead time in production planning and production ordering

In the case of "Push System", production planning for multi-stage production processes is generally based on D_t, $t-\Sigma LT^k$. (Fig. 8)

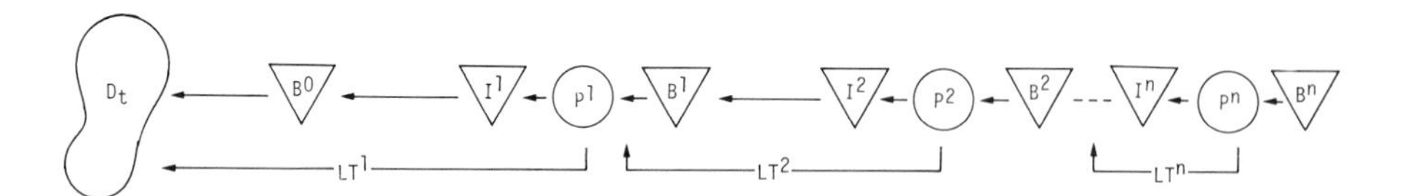

Fig. 8 Production Process and Lead Time

That is, production planning for a certain process is based on the demand forecast at the time prior to the aggregate lead time between that process and the final process. Then, the production planning is given to each process as a production order.

In the case of "Pull System", production planning for multi-stage production system is based on $D_{t}, {}_{t-LT^{1}}$. That is, production planning is made for only the final process, and it is distributed to only the final process P^{1} as a production order. Basically production orders for the other processes are given automatically through Kanban System. What enables us to conduct Kanban System is smoothing of production at the process P^{1}.

The difference in the lead time between "Push System" and "Pull System" results in the difference in the forecast error in the demand, as we mentioned before. The difference of the production ordering system in all processes preceding the final process results in the difference in their lead times changing production ordering when the process P^{1} requires them to do so.

We, at Toyota, place great importance on especially shortening the information processing lead time in LT^{1}. To be more specific, it means how short the lead time from the customers' order receipt to production ordering is. In such information processing stage, we of course utilize electronic computers to develop a smoothed production planning, which, we consider, is also a sort of CAM.

5. CONCLUSION

The automotive industry and the market has become matured, which implies that the day of a severe competition has come. In such an environment, it is necessary for the auto manufacturers to respond to various changes. We first introduced the idea of FMS in machine shops more than 20 years ago. We have gradually expanded application of such a system, with the result that Toyota and the vendors now conduct Toyota Production System on the basis of Just-in-time. However, we still have some problems which hinders the progress of flexible production system. We would like to solve them referring to studies and researches in various fields.

REFERENCES

1. Kimura, O., and Terada, H. Design and Analysis of Pull System, a method of multistage production control. INT. J. PROD. RES., 1981, VOL. 19, No.3, 241-253.

2. Muramatsu, R. The Theory and Practice of Production Management. Journal of Japan Industrial Management Association, 1980, VOL. 30, No.4.

3. Sugimori, Y., Kusunoki, K., Cho, F., Uchikawa, S., Toyota Production System and Kanban System - materialization of just-in-time and respect for human system. PRE-PRINT, 4th International Conference on Production Research, 1977.

FLEXIBILITY AND PRODUCTIVITY IN COMPLEX PRODUCTION PROCESSES

Sten-Olof Gustavsson
Department of Industrial Management
Chalmers University of Technology
Sweden

ABSTRACT

The drastical changes in market demands, and the rapid technological development, has created a need for:

- more flexible production systems
- more complex products with a larger degree of variation.

There is a strong force towards the use of more and more mechanized and automatized equipment, from single NC-machines to complete manufacturing systems. At the same time there is a need for flexibility towards changes of the products. These changes have to be made in a limited time, and without the need of large reinvestments in the production system.

This means that there more often must be a discussion regarding flexibility versus productivity before the production system is designed. I will discuss

- methods for calculation of different flexibility levels.
- strategies for a more flexible view upon products and processes.
- examples and results from different areas within the Swedish industry.

INTRODUCTION

As a result of increased industrial automation and of the trend towards an ever shorter life cycle for a product, it has become apparent that the flexibility of the machinery needed for complex production processes is now of overriding importance for long-term profitability.

The danger exists that a short-term gain in production is achieved by using machinery or equipment that then becomes redundant on the introduction of a new model. There might thus be a conflict of aims between flexibility and productivity. Strategically speaking, production should be so flexible that neither the product nor the renewal of the processes should be hindered by "sunk" costs in production.

Productivity

This concept is familiar and has been dealt with in a variety of contexts so only a few aspects will be touched on here very briefly.

Productivity

- corresponds well with mass production (over a long period)
- brings the short-term perspective into focus
- draws attention to internal (the production apparatus) rather than to external questions (what the client judges as valid i.e. the right product at the right price).

Albeit simple in theory, the concept of productivity is beset by complications and difficulties in practice since how it may be interpreted will depend on factors such as the time aspect, product development, inflation and econometrics.

Considerable advantages can be gained by utilizing straightforward simple measures as work productivity in physical measures of quantity per employee, output productivity and capital productivity without coupling this to the total productivity index.

Key figure comparisons with competitors will help to give a very comprehensive view of the extent and power of the competition. Obviously, in every industry, there must be a continuous follow-up of production, preferably with the help of several key figures. Assessment, for example, of a 10 percent productivity increase should be a familiar routine in every business.

It is equally obvious that continuous comparison should be made with other manufacturers within or outside the company. The real challenges, however, arise every time a decision is made for the future, which is to say, that action should be taken on the basis of accumulated experience. The investments in all these resources are expected to pay off (at least in the long-term). This means that the use of these resources will often make demands on both productivity and flexibility, and once more, the needs of the market are decisive.

Productivity and flexibility.

Production (i.e. goods) is defined as the manufacture of products with the help of personel, material, equipment (hard and soft-ware) and capital. The consumption of resources is compared with earlier consumption in budget control and other steering instruments.

Products are subject to changes:

- a change of technology (electronics take over from mechanics),
- "rationalisation" (one component does the work of several),
- changes in fashion

A company's ultimate success depends on its ability to utilize resources and meet the needs of the market. These internal factors steer demand and in turn the volume of business and the price of the commodity.

In addition to all this, there must be flexibility in respect of external factors.

These may be:
- fluctuations of the market,
- seasonal fluctuations,
- competition from other companies.

There is also another interesting dimension to take into account, namely the question of the product's life cycle. Uncertainty is always present from the moment a product is introduced on the market (will it be a success or not?) to the end (when will the market vanish?).

Flexibility

Flexibility comes from the Latin word for bendable. Other expressions are adjustable and mobile. Industrially speaking the word means adaptable and capable of change. The concept has been a subject of interest to both production engineers and research workers. The flexibility of the work group has been examined by, for example, Kozan (1982) and the flexibility of the manufacturing system by Hjelm (1982). Warnecke et al (1981) have discussed the flexibility of the whole production system.

Flexibility can be defined as follows:

1. Changes in the product
   - improvements, new components,
   - several variants
2. Changes in the production system
   - new machinery and production methods,
   - new systems (for example, computerization),
   - new personnel

3. Changes in demand
   - insecurity over a period,
   - fluctuations (over the year, for example).

It can be said that these three types of changes made demands on the production system as regards both the short-and long-term view. I can make the following basic divisions according to the time aspect involved when it comes to assessing productivity versus flexibility.

1. Operational problems. Short-term - for example, having to replan in order to cope with a breakdown of vital machinery, or an unexpected shortage of material.
2. Tactical problems. Medium-term, such as changes in design or rate of production.
3. Strategic problems. Decisions with long-term effect such as investments in machinery or expansion.

I have three examples to illustrate the differences

A. Shipyard
   Ships - alternative production
B. Hobby equipment
   Lawn mowers - snow scooters
C. Car factory
   Line-out

It is essential to identify "the critical time perspective or perspectives", that is, to be ready to cope with the operationel, tactical and/or strategic problems. This analysis, of course, should be made preferably when the system is constructed but since external circumstances constantly alter, there should be discussion of these matters at regular intervals.

If the analysis is to be valid it must look closely at not only the production system's resources (personnel, machinery, etc) but also at the qualities of the product or products themselves. Which of the above is crucial for success? The answer would seem obvious: the product. No customer - no business, which means that a thoroughly attractive product is the be-all and end-all of the matter. However, attractiveness is not just a question of function, it is also one of price and quality. There is thus a clear connection between the product and the production system with its resources.

In conclusion, I would like to point out that in every production system the following must be decided:

1. which level is primarily critical (for the company and sections of the company) operational, strategic, tactical,

2. which resource (personnel, machinery/system or product) is primarily critical and therefore in need of particular attention.

It is only after the crucial factors have been identified that the work of constructing an effective (productive and flexible) production system can begin.

I should like to illustrate our reasoning with another example which is interesting in that it demonstrates all three types of flexibility seen from the strategic, tactical and operational angle.

Example: Truck manufacture.

A case study of this type can always be said to be unique but one should bear in mind that:

A. Mass production of exactly the same items is not so common but products serve the same purpose, the only difference between them being the design of a few details.

B. Variations of the same basic construction can be designed in different ways.

The list could be made longer. It is enough to say that all production systems are unique, some more unique than others. Naturally, a concept such as flexibility (or productivity) cannot be expected to follow a simple standard pattern. Let us instead indicate approaches to the problem.

F	Capacity	- Change of volume
L		
E	Product	- Design, models, generations
X		
I	Steering system	- Structural programme, raw
B		material/primary products
I	Production	- Direction, flow
L		
I	Machinery	- Machines, tools, fixtures
T		
Y	Personnel	- Competence, structure

Fig. 1 Structures in the concept of flexibility

Good examples of flexible equipment are the NC-Machines which have reduced the rigging or starting-up time for machines. This has allowed for economical small-scale production where previously multi-specialized operations in several installations required large-scale series for manufacture to be economically viable. A good example of product flexibility is a construction composed of modules whose end-product, although made up of a unique combination of modules, is itself composed of mass-produced components. A pizzeria is another example.

Good flexible capacity can be achieved by utilizing the parallel principle, whereby a parallel product line can be added or discontinued as demand varies. Good examples, by all means, but it must be possible to measure how good they are. To do this requires inventiveness. Here are a couple examples:

1. Flexibility of machinery can be measured as the ratio of the investment's residual value for the next product model to the original investment, i.e. an index between O and 1.

2. Product flexibility can be measured as the ratio of the residual value of the old model to the new model divided by the original value for the old model.

Strategies for the assessment of flexibility versus productivity

There are at least three ways of calculating so-called "optimal" flexibility.

I All starting-up costs and all other costs are grouped under "life-cycle cost" to facilitate optimization.

II Only model-restricted machinery is optimized in the "life cycle" sense

III Expensive process machinery is made to be used irrespective of model and has been standardized to the extent that it can be used generally. To take an example: the Swedish Match Company by the 1920's had already standardized its matchstick to the point that the company could risk building fully automated machinery for mass production of the matchstick at the same time that it could offer a large assortment of shapes and sizes for the boxes and labels.

Model 1 has been applied in the aviation industry where so-called "break-even" calculations have been made including all costs; product development, prototypes, testing, grounds, buildings, tools jigs, fixtures etc. An estimate is then made of the size of the prospective market, for example 300 planes. It is then a simple matter to evaluate the result of bringing in more mechanized and automated equipment. A saving of 1000 Swedish crowns in working costs per plane is the equivalent of a maximum investment of 300,000 crowns.

In principle, the model is simple but the limit for the total length of the series must be set so low that it is certain to be exceeded and the model gives priority to alternatives bringing in a profit.

Model II is applied in the motor/automotive industry, where costs are separated into model-restricted and model-free categories. If all doubtful costs are classified as model-restricted, it is possible even here to calculate the value of, for example, work-saving automated machinery. A saving of 100 Swedish crowns in work costs is the equivalent of a maximum investment of 100 million crowns for a total life cycle length of 1 million cars.

In the April 5th, 1982 issue of "Fortune" Porsche's car body assembly was described as flexible, manual with 25 man-hours per car. A more normal plant for a series length of 250,000 cars a year would require 5 man-hours per car. Since Porsche only produces about 10,000 of the 911-model per year, the difference would mean a capital cost of 2,000 Swedish crowns per car, a total of 20 million crowns a year. A possible investment of roughly 80 million crowns in all for mechanized machinery. This is probably not feasible at this cost. If one reckons on a volume of 250,000 cars per year, the value of 20 man-hours saved is roughly 500 million crowns per year and a total possible investment of 2,000 million crowns, which could be done with good profit. The conclusion is that it is easier to achieve flexibility with small volumes for economic reasons.

These examples shows that rough estimates can be made which will serve for the assessment of different production systems.Unfortunately, this seldom happens, possibly owing to a widespread belief that the higher the degree of flexibility the lower the level of productivity. (cf. Warnecke et al. 1981).

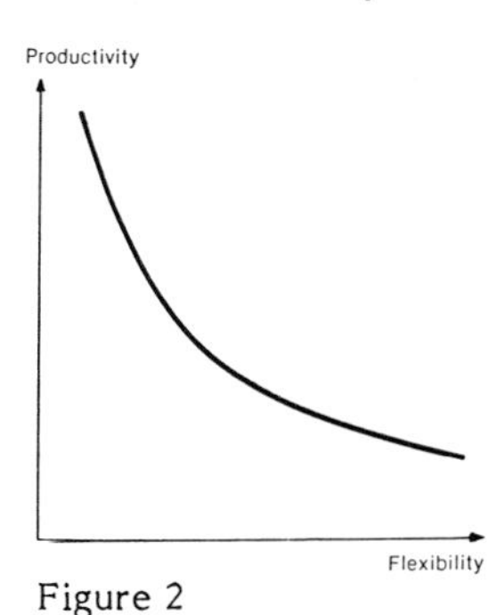

Figure 2

It is also commonly believed that increased flexibility must involve higher investment costs.

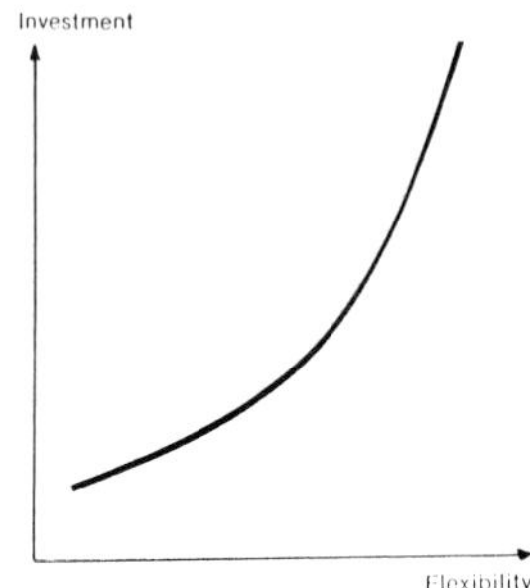

Figure 3

The relationship between flexibility and efficiency is heavily dependent on the structure of the chosen system and so the design of the system becomes instrumental to the success or failure of the relationship.

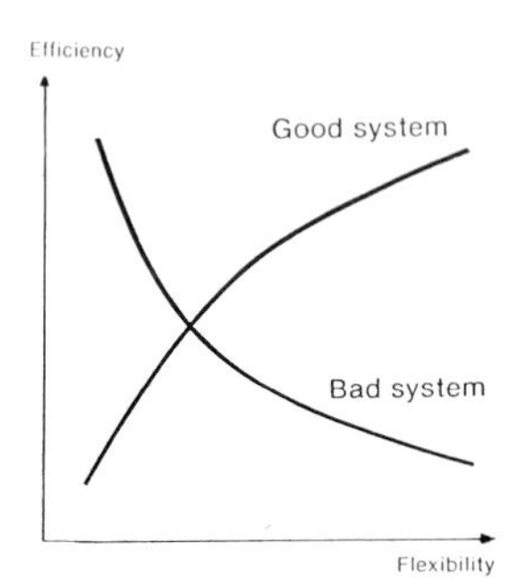

Figure 4

Which strategy for flexibility?

Demand flexibility

If all the analyses indicate that there is little risk of significant changes in demand, then investment can be made in a process which constitutes a total solution for a fixed capacity. Generous margins will ensure a certain amount of flexibility. However, if there is a risk that demand will vary significantly, and if the total demand is hard to estimate, then investments should be made in small parallel steps which can easily be added to or stopped short.

Product flexibility

If there is a likelihood that there will be frequent changes of model and product, then the product should be composed of modules with allowance made for step-by-step changes. It is, of course, risky to invest in module-restricted process machinery for suchlike products.

Flexibility of machinery

If it is very likely that there will be a technological change from, for example, steel to plastic, then it is advisable to employ subcontractors who have several other customers with other products which facilitate the technological transfer.

SUMMARY

Conclusions

A consciously planned flexibility can be achieved without significant increases in investment. The following guide lines can serve for the design of both product and prodution systems.

The use of modules

Many alternative machinery installations can be converted if each module has access to air, electricity and compressed air, for example. Standard components can be combined as larger pieces of mechanical equipment. Products can be renewed successively if they are composed of modules. ASEA, for example, has constructed switch gear by using modules.

Standardization

Well-planned rigorous standardization will facilitate the introduction of larger series, improve economy and encourage demand. The Hasselblad system, for example, was constructed with a number of rigorously standardized measurements which can be retained more than 30 years later.

The Variant tree

Intitial operations in a process should not include variants. These, however, can be produced economically if the final operations encompass a large number of alternatives.

Development towards automation

The requirement that all automated machinery can be quickly adjusted to new dimensions promotes flexible machinery. Rapid changes in the machinery allow for small series and foster a large capital turnover. Life-cycle costs have become increasingly relevant to the decision-making process provided that the aim is to review the cost of the whole life-cycle for a product and/or production system. It is necessary, however, to include appraisals of the development of volume, inflation and shifts of balance in the cost relations.

Last but not least, there is a need for efficient personnel at all levels. This means they should be both productive (interested in making improvements and flexible (able to revise their thinking). Every manager's primary task is to create such personnel.

ACKNOWLEDGEMENTS

This paper has been prepared in collaboration with my close friend Bengt Almgren to whom I wish to express my gratitude.

REFERENCES

B Colding Productivity gains through the realization of the integrated manufacturing system today in limited manpower production at low cost.

S. Hjelm: Flexibla Automatiserade tillverkningssystem - Mål och förutsättningar i Sverige. The Royal Institute of Technology, Stockholm, Sweden, 1982

K. Kozan: Work Group Flexibility: Development and Construct Validation of a Measure. Human Relations, Vol 35, nr 3, 1982, pp 239-240

H-J Warnecke, H-J Bullinger, J.H. Kölle, German manufacturing industry approaches to increasing flexibility and productivity. World productivity conference Detroit 1981.

S-O Gustavsson, Motive forces for and consequences of different plant size VI International conference on production research Novi Sad 1981.
Fortune, April 5, 1982 Automaking on a human scale. p 87-93.

PROBLEMS, SOLUTIONS AND EXAMPLES OF INDUSTRIAL APPLICATION OF ROBOTICS

Prof. Dr.-Ing. H. J. Warnecke

Fraunhofer Institute for Manufacturing Engineering and Automation, IPA, Nobelstr. 12, 7000 Stuttgart 80, Federal Republic of Germany[x)]

x) Dipl.-Ing. E. M. Wolf, research fellow at IPA, collected the information presented in this paper

ABSTRACT

An overview of the areas in the production process where robots are used in the German Federal Republic as well as the numbers of installations are presented. Two examples for industrial applications of robots, problems and necessary developments are explained. These examples concern the automatic assembly and the automatic deburring and fettling, two fields where robot applications will increase considerably in the next 10 years.

1 INTRODUCTION

Industrial Robots are known in industry for about 20 years with the first application on the shop-floor not earlier than in the first 70´ies. Today about 7000 Industrial Robots exist in Europe from which 50 % are used in the Automobile Industry.

Already in 1974 more than 1000 Industrial Robots were installed in the US and especially in Japan. In Europe on the other hand robots were applied rather hesitantly. Fig. 1 shows how rapidly the robot population is rising in Japan from 1978 to 1982 reaching 12000 robots at present time. In these four years the number of robots in the US increased less rapidly and by now in Europe about the same number of Industrial Robots is installed.

2 FIELDS OF ROBOT APPLICATION IN THE GERMAN FEDERAL REPUBLIC

At present almost 4000 Industrial Robots are used in Germany. The main field of robot application and the number of industrial applications are shown in Fig. 2.

Besides in the classic fields of robot application such as spot-welding, arc-welding, spray-painting and machine loading, high increases in robot population are expected in the wide field of assembly and in fettling and deburring. For each of these two promising fields one example of robot application will be presented in this contribution. Today, approximately 80 Industrial Robots are installed in the field of fettling, deburring and brushing. Many casting processes cause burrs with different shapes and sizes. Therefore the cleaning of castings is a common process in the foundry industry. The size and form of the burrs to be removed depends considerably on the applied casting technology and the quality of the molds.

With new developments on the field of sensors, adaptive controls and improved programming methods, robot application can be increased in this area considerably.

For assembly operations 122 Industrial Robots are used. According to a Delphie-Study conducted by the Fraunhofer-Institute for Production Engineering and Automation, in no more than 2 years it will be possible to assemble several different products per shift and in 1985 the robotic assembly of car-aggregates will be state of art.

3 APPLICATION OF ROBOTS FOR FETTLING AND DEBURRING

In order to increase the number of robot applications in this area, new developments are required in the field of sensors, adaptive controls and programming methods.

For these reasons, a fettling laboratory was set up at the IPA in Stuttgart. Right now, this laboratory is equipped with 3 different types of robots (KUKA 601/60, Hitachi Process Robot, Puma 600) and a considerable number of different fettling tools and sensors. With this equipment, it was possible to test different adaptive control strategies and programming methods.

3.1 Sensors

If there are high demands regarding the smoothness of the surface, geometrical measuring quantities will provide for better results. Fig. 3 shows the possible geometrical measuring quantities and different concepts for sensor systems. Tactile sensors for scanning a line or a matrix were developed and tested in different research projects /1/. This type of sensor can monitor the complex contours during the grinding process.

3.2 Robot Workplace to Grind Wooden Parts

The application of Industrial Robots for the grinding of wooden parts was demonstrated in the laboratory mentioned earlier. The demands for this laboratory set-up concerned wooden parts for the furniture industry, especially intricately formed legs of tables and chairs. The geometrical shape of these parts is produced by a milling machine (copy milling or form milling). After milling, to achieve the desired faultless surface, the wooden parts must be grinded, this was hitherto performed, completely manually, in two work cycles (rough grinding and fine grinding). So far, any attempts to automate this procedure have failed, among other things, because of the lack of flexibility and the small number of parts. Tests have been carried out at the IPA, in which an industrial robot takes the wooden parts from a magazine with a special gripper to pass them on to the grinding tool for the grinding procedure. After the grinding procedure the wooden parts are deposited in another magazine (Fig. 4).

Since as much of the surface of the wooden parts as possible must be grinded, the industrial robot gripper must clamp the wooden part in a very small area. Because the wooden parts were very large, rather large clamping tolerances occurred when the small gripper was used. To achieve a good surface quality, there had to be picked up and compensated for with the aid of the sensor.

Since the geometrical shape of the wooden parts was predetermined, it seemed more appropriate to apply a technological, rather than a geometrical sensor. The power consumption of the grinding tool was selected as the measuring value.

The laboratory set-up was largely made up of the following components:

- o special tool for wood grinding
- o Industrial robot KUKA IR 601/60 CP
- o measuring device for continuous control of grinding force
- o processing of measured values to adapt the measuring signal to the industrial robot control
- o sensor interface for the industrial robot.

Aided by the sensor interface, correction data for the alteration of the fixed programmed path may be given to the industrial robot.

During the test runs it was found that such an adaptive control may be realized. As soon as the wooden part was pressed too strongly against the grinding tool because of clamping tolerances, the grinding capacity exceeded its minimal value. Thereby, the programmed path of the industrial robot was automatically corrected (away from the grinding tool). If, however, the capacity remained under a particular minimum value, a correction of the programmed path into the direction of the grinding tool automatically followed. In the investigation two contours of the wooden parts lying opposite each other were grinded. The consequence of clamping errors was that too much was grinded on one side and too little on the other side. By the application of the sensor system described, an approximately even grinding force on both sides despite of clamping tolerances could be achieved.

The investigations proved that such tasks may be solved today. Problems occur in the computing time necessary for the processing of measured values and in transferring the measured results to the industrial robot control. The correction of stored programmes has hitherto generally required a large amount of effort in order to modify that hardware and software of the industrial robot. Since more and more robots are now equipped with standard sensor interfaces for the correction of programmes, the use and application of the described sensors will surely be simplified quite considerably during the next few years.

4 PROGRAMMABLE ASSEMBLY STATION FOR THE AUTOMATIC ASSEMBLY OF CAR AGGREGATES

The Fraunhofer-Institute for Manufacturing and Automation (IPA) has developed a Programmable Assembly Cell to assemble a large variety of automotive parts and units. Those operations have been performed manually or by single purpose automatic machines so far.

Fig. 5 shows how this Assembly Cell, which was presented on the Hannover Industrial Exhibition can be integrated in an existing assembly line with the workpiece on free flowing pallets. The line may have manual as well as conventional automatic stations.

The Programmable Assembly Station features two industrial robots. One is a handling robot to grip and position parts (see Fig. 6) and the other one is fastening bolts using a D.C. electric nut-runner with torque/turning angle control and an automatic system to change sockets (see Fig. 7). The tightening control of the nut-runner is programmable.

The gripper of the handling robot has parallel jaws with a programmable stroke and a programmable gripping force. Two sensors are measuring the gripping and the joining force.

Parts to be assembled are presented by feeding devices or magazined on plastic pallets. The pallets are moved into the working space of the handling robot out of a storage system to ensure a sufficient reserve of parts for extended unmanned operation periods.

At the work place for the second industrial robot screws must be presented to the nut-runner. Depending on the type and number of different screws to be used, several screw magazines, vibratory feeder bowls and similar devices have to be provided.

The feeding of the screws to the screwing mechanism may either take place directly via air pressure activated feeding lines, or the robot picks up each screw individually by means of the screwing mechanism from a feeding device. For tightening different types of screws it is necessary that the robot handles several screwing tools. The robot changes these tools automatically with the aid of a change mechanism.

For conventional assembly lines the multi-screw-heads of the automatic stations commonly used are in fact only utilized to a very low percentage. This is being caused by the fact that the cycle time is tuned to manual assembly. During the entire cycle time each spindle of the multi-screw-screwing head only has to perform one screwing operation. When inserting screws by means of industrial robots simpler screwing mechanisms can be used, which during one single cycle time perform several screwing operations and are thus better utilized.

As a result of the considerable work content of the industrial robot, which may be adjusted to any individual assembly sequence, these devices are well utilized. They are able to assemble different variants and it is possible with little effort to assemble other types in these stations. Because of the complex joining movements possible with industrial robots, a similar application in assembly with fixed automatic stations would not be possible. The workpiece spectrum to be assembled in the pilot working place of the programmable assembly cell is shown in Fig. 8.

5 FUTURE TRENDS

The development of sensor techniques and robot control systems which can be interfaced with sensor-signals will open new fields of application as well as increase the flexibility of existing application possibilities.

Despite the high expectations to sensor techniques it must be mentioned that there already are and there will be numerous robot applications without sensors.

Considering social aspects, improved working conditions as well as the changing qualification of the work force, it is certain that in the late 80'ies industrial robots will be as common in the production process as NC-machines are today.

Industrial robots should not be considered as the solution of any automation problem but in many fields in the production process the industrial robot became state of the art and is expected in the German Federal Republic to strengthen the position on international markets.

6 REFERENCES

/1/ Stute, G.; Erne, H.:
The control design of an industrial robot with advanced tactile sensitivity. Proc. of the 9th ISIR. Washington, 1979.

/2/ Bäßler, R.; Schunter, J.; Spingler, J.; Walther, J.:
Einsatz von Industrie-Robotern in der PKW-Aggregatemontage. Proc. of the 3rd Assembly Automation, 14. IPA-Arbeitstagung, Böblingen, Mai 1982.

/3/ Warnecke, H.-J.; Schraft, R.-D.; Schweizer, M.; Walther, J.:
Applications of industrial robots for assembly-operations in the automative industry. Proc. of the 13th ISIR, April 1983, Chicago, USA.
Conference Proc., Dearborn: SME, Vol.1, pp. 5-30 - 5-39, 1983.

/4/ Warnecke, H.-J.; Walther, J.:
Programmable assembly cell for automative parts and units. Proc. of the 2nd Europ. Conf. on Automated Manufacturing. pp. 159 - 169. Birmingham: 16. - 18. May 1983.
Ed.: Rooks, B. W.: Kempston: IFS Publ.
Amsterdam, U. A.: North Holland Publications, 1983.

/5/ Schraft, R.-D.; Schweizer, M.; Abele, E.; Sturz, W.:
Application of Sensor Controlled Robots for Fettling of Castings. Proc. of the 13th ISIR, April 1983, Chicago, USA.
Conference Proc., Dearborn: SME, Vol. 2, pp. 13-44 - 13-57, 1983.

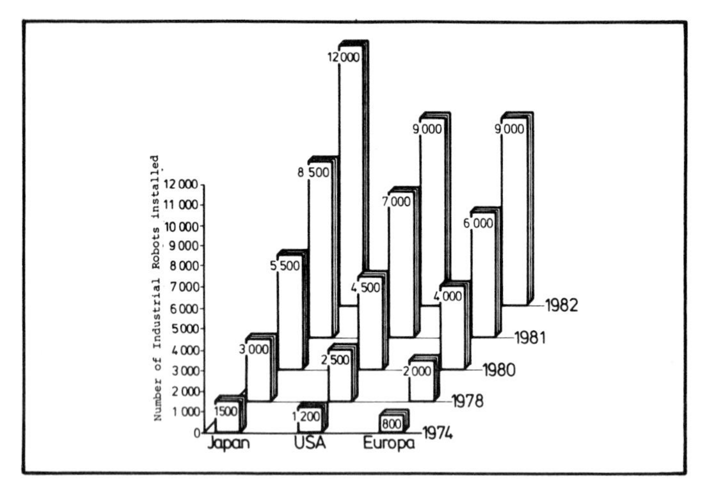

Figure 1 : Installations of Industrial Robots in Japan, US and Europe

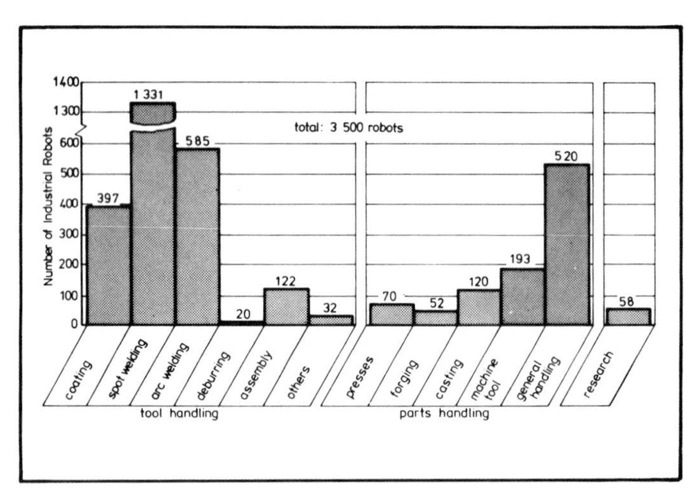

Figure 2 : Application-fields of Industrial Robots in the German Federal Republic

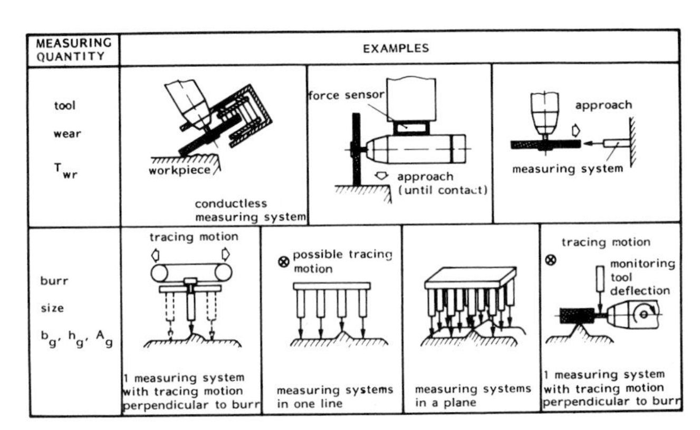

Figure 3 : Measuring quantity and measuring principles of sensors for geometry

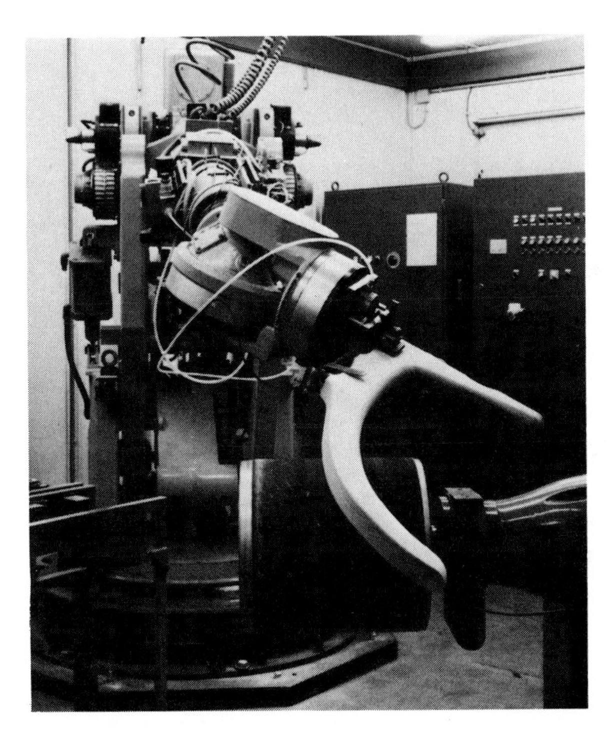

Figure 4 : Grinding of wooden parts with an industrial robot

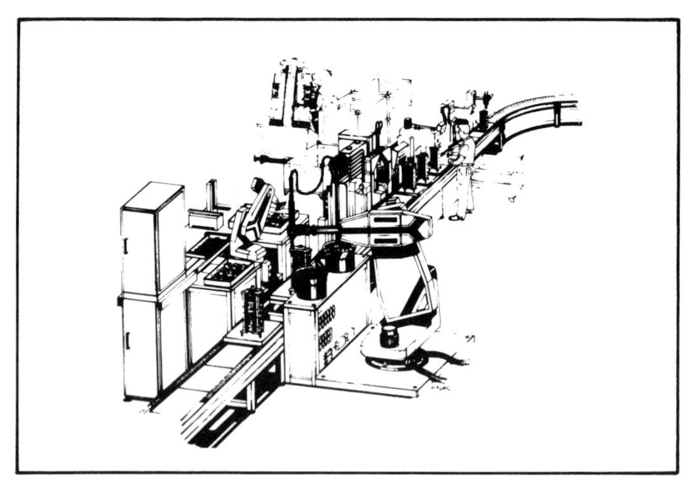

Figure 5 : Assembly line for car aggregates with integrated Programmable Assembly Station

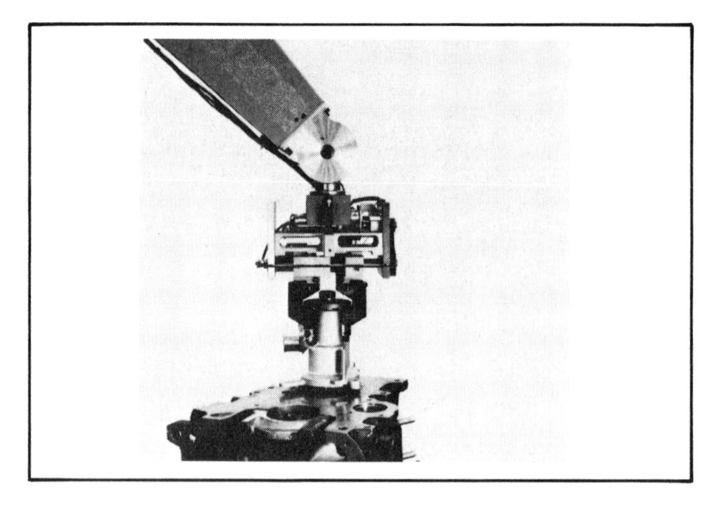

Figure 6 : Handling robot with gripper

Figure 7 : Robothand with electric nutrunner

Figure 8 : Parts of a motor to be assembled in the Programmable Assembly Station

THE MACHINE INTERFERENCE PROBLEM IN MANUFACTURING CELLS WITH INDUSTRIAL ROBOTS

E.A. ELSAYED

Department of Industrial Engineering, Rutgers University, P.O. Box 909, Piscataway, N.J. 08854 (USA)

ABSTRACT

This paper investigates the machine interference problem in manufacturing cells where one robot has charge of a number of identical or similar machines. In this paper, we determine the optimal number of machines (each is liable to two types of failures) to be assigned to the robot (the robot is also subject to failure) such that the total cost of the manufacturing cell is minimized. Two repair (service) policies are compared. Machine availability and robot utilizations are also obtained.

INTRODUCTION

The introduction of flexible manufacturing systems and programmable robots is one of the main factors which resulted in the increasing trend of using independent manufacturing cells connected through a reliable network of material handling systems over the widely automatic flow lines (Fig. 1).

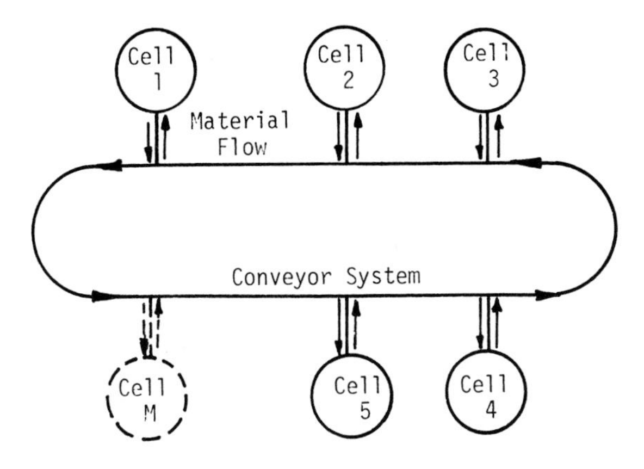

Fig. 1. Independent manufacturing cells in a typical flexible manufacturing system.

In fact, there are situations where a flow line can be replaced entirely by several manufacturing cells connectec by a materials handling system. Elsayed and Norton (ref.1) suggest the replacement of a flow line for the manufacturing of cross deck pendant terminals by a manufacturing cell served by a single robot. This leads us to the definition of a manufacturing cell. It is the smallest section of manufacturing consisting of one or several CNC-machine tools, an industrial robot and a workpiece storage, in which the processes of machining, handling and control take place automatically as shown in Figure 2.

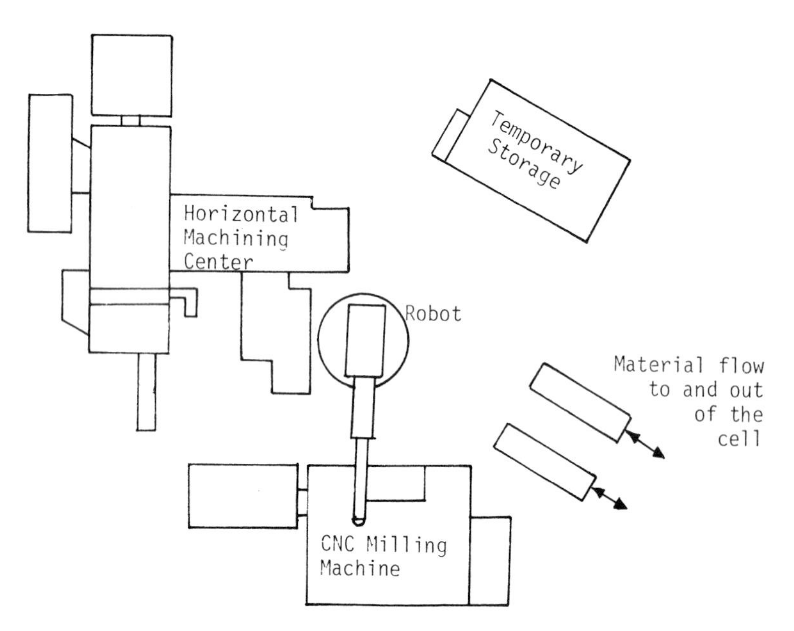

Fig. 2. A typical manufacturing cell serviced by a single robot

The robot (operator) becomes active only during the set-up, loading, unloading and maintenance. In such cases the automatic process in the manufacturing cell is interrupted either partially or totally (Schmigalla and Rodeck (ref.2)).

In a typical manufacturing cell, one robot attends a number of identical or nonidentical machines. From time to time a machine stops and does not resume production until it has been serviced by the robot. If two or more machines are stopped, only one machine can be attended at once, and so production is lost while machines are stopped awaiting attention. The rates of production per machine and per robot thus depend not only on the rate of occurrence of stops and on the "attention-time" per stop, but also on the number of machines in the manufacturing cell. This problem is usually referred to as *the machine interference problem*.

Through the years, the problem of machine interference has been studied by many researchers. Stecke (ref.3) presents a comprehensive review of the operator/machine interference problem. She (ref.4) also presents a complete summary of the many approaches that have been used to deal with this problem. These approaches are classified as: deterministic, probabilistic and simulation approaches. We now review the research that used these approaches.

Virtually all of the deterministic models determine the number of machines assigned to a single operator based on the minimum sum of the costs of idle labor and idle machines. It is assumed that the processing times of the machines, loading and unloading times as well as both the setup and service time are deterministic and exactly known. The most used method for finding the assignment under the above conditions has been through the use of man-machine charts. The charts, however, are only applicable to small scale problems, provide no measurement of interference level experienced, and do not always provide the optimal solution.

One of the earliest attempts in measuring the interference in a system is presented by Wright (ref.5). He assumes that failures of machines are inversely proportional to the number of functioning machines at any given time.

A branch and bound algorithm for application to machine interference problems is presented by Miller and Berry (ref.6). No transit time of the operators or preempting of service is permitted and all operators and machines are assumed indentical. Dube and Elsayed (ref.7) propose a more general model than Miller and Berry (ref.6). They (ref.7) assume that processing times are known and deterministic while service times are normally distributed.

All deterministic models do not include any random occurrences of stoppage of machines or service by the operator, which makes those models inappropriate to use in many situations.

Therefore, researchers have focused their investigation to model the machine interference problem as a probabilistic problem and approached its solution by using either queueing theory or simulation approaches. We now present a summary of the research which utilize these two approaches.

Early queueing models are presented by Palm (ref.8), Ashcroft (ref.9), and Benson and Cox (ref.10). Without finding a general solution, Benson and Cox propose a model for machines subject to two types of failures. Jaiswal and Thiruvengadam (ref.11) extend Benson and Cox's model (ref.10) allowing each machine to be liable to two types of failure while the repair times follow a general distribution.

Hodgson and Hebble (ref.12) investigate a non-preemptive priorities model where one operator tends K batteries, each having some finite number of identical machines. Reynolds (ref.13) investigates the shortest distance priority service discipline. He states that "The difference between FCFS and the shortest time distance service disciplines becomes insignificant as the ratio of machines to repairmen becomes large." Elsayed (ref.14) proposes two repair policies for the machine interference problem when machines are liable to two types of failures. In the first repair policy of Elsayed (ref.14) one type of failure is assumed to have the "head-of-the-line" priority over the other type, while in the second policy failures are repaired with equal priority.

Using the simulation approach, Reich (ref.15) extends Benson and Cox's work (ref.10) to determine the machine utilization of a man-machine system when machines are subject to two types of failures. Haagensen (ref.16) develops a general simulation model to determine the optimal assignment of machines to operators. Freeman, et.al. (ref.17) examine a staffing problem for tending a multiple machine system. One type of stoppage is permitted in the model and the FCFS and the Shortest Processing Time (SPT) discipline are compared.

Garcia-Diaz, et. al (ref .18) combine the simulation approach with direct search optimization techniques for allocating heterogeneous atuomatic machines to workers. Guild and Hartnett (ref.19) extend Reich's model by simulating a more complex two-stoppage interference model with N operators and M machines. They find that the complex system could be approximated by a one-stoppage analytical model with N operators and M machines.

Medeiros and Sadowski (ref.20) provide a general modeling approach for the design and analysis of computer-controlled manufacturing cells containing robots, sensors, automatic machines, and orientation devices. They utilize Q-GERT for the simulation of the problem. Karmeshu and Jaiswal (ref.21) consider a non-linear stochastic model for the single-server machine interference problem where the repair rate is a rational function of failed machines. Elsayed and Dhillon (ref.22) present steady-state and transient solutions for the machine interference problem when the service rate is a function of the number of failed machines.

Recently, Lehtonen (ref.23) considers an exponential repair model with s machines and one repairman. He assumes that the machines' failure rates are equal but the repair rate may change from one machine to another.

The solutions presented by most authors are based on the assumption that the operator is always available for servicing (repairing) the machines.

The purpose of this paper is to analyze the machine interference problem in manufacturing cells serviced by an industrial robot. It is assumed that the robot is subject to random failures. It is intended to determine the optimal number of machines (each machine is liable to two types of failures) to be assigned to the robot under two different service policies for the machine failures such that the total cost of the manufacturing cell is minimized.

NOTATIONS

We first present notations which are general and used in the two repair (service) policies of the machines, these notations are:

M	Total number of machines in the manufacturing cell
λ_i, μ_i	Constant calling or failure (hazard) and repair (hazard) rates for failure type i (i=1,2), respectively
λ_s, μ_s	Constant failure (hazard) and repair (hazard) rates of the industrial robot
ρ_k	Traffic intensity of failure k ($\rho_k = \lambda_k / \mu_k$)
N*	Optimum number of machines assigned to the robot
M.A.	Machine availability
R.E.	Efficiency of the robot (robot utilization)
R	Fixed cost rate of the robot
F	Net profit rate of each machine

The following notations are used for the first service (repair) policy (POLICY A)

$P_k(o,y,z)$	Probability that there are y and z failed (require service) machines of types 1 and 2 respectively while the robot is operating and the machine failure (service) being repaired (performed) is of type k (k=1,2)
$P(1,y,z)$	Probability that there are y and z failed (stopped) machines of types 1 and 2 respectively awaiting to receive service from the robot while the robot is under repair
$E[N_k]$	Expected number of machines not in working condition owing to failure type (or service call) k

The notations for the second service (repair policy (POLICY B) are

$P(o,y,z)$	Probability that there are y and z machines awaiting to receive service (repair) from the robot which is busy servicing another machine
$P(1,y,z)$	Probability that there are y and z failed (stopped) machines of types 1 and 2 respectively awaiting to receive service from the robot while the robot is under repair
$E[N]$	Expected number of machines not in working condition

ASSUMPTIONS

1. Each manufacturing cell has M identical (or similar) machines; each machine is independently liable to two types of failures (types 1 and 2) or may be calling for two types of service of the robot.
2. The time for service calling (or failure) and time of service (repair) are exponentially distributed with parameters λ_i and μ_i (i=1,2). While the time to failure and the time for repair of the robot are exponentially distributed with parameters λ_s and μ_s respectively.
3. The robot is liable to failure regardless if it is servicing a machine or not.

4. We compare two repair (service) policies of servicing the machines. In Policy A, the robot services the machines according to the priority assigned to the failure (the robot may not interrupt the service of a failure of a lower priority to serve a higher priority failure). In Policy B, the robot serves the machines with equal probability.
5. The probability of simultaneous events in infinitesimal intervals of time is zero.

THE MODELS

We now describe the two repair policies.

1. POLICY A:

The steady-state probability equations which describe the manufacturing cell under consideration when priority is placed on one type of failure over the other are given below:

$$[M(\lambda_1+\lambda_2)+\lambda_s]P(0,0,0) = \mu_2 P_2(0,0,1) + \mu_1 P_1(0,1,0) + \mu_s P(1,0,0) \tag{1}$$

$$\begin{aligned}[(M-i)(\lambda_1+\lambda_2)+\mu_1+\lambda_s]P_1(0,i,0) &= [(M-i+1)\lambda_1]P_1(0,i-1,0) \\ &+ \mu_s P(1,i,0) + \mu_2 P_2(0,i,1) \\ &+ \mu_1 P_1(0,i+1,0) \qquad i=1,2,\ldots,M-1\end{aligned} \tag{2}$$

$$[\mu_1+\lambda_s]P_1(0,M,0) = \lambda_1 P_1(0,M-1,0)+\mu_s P(1,M,0) \tag{3}$$

$$\begin{aligned}[(M-j)(\lambda_1+\lambda_2)+\lambda_s+\mu_2]P_2(0,0,j) &= [(M-j+1)\lambda_2]P_2(0,0,j-1) \\ &+ \mu_1 P_1(0,1,j) + \mu_s P(1,0,j) \\ &+ \mu_2 P_2(0,0,j+1) \qquad j=1,2,\ldots,M-1\end{aligned} \tag{4}$$

$$(\mu_2+ \lambda_s)P(0,0,M) = \lambda_2 P(1,0,M-1)+\mu_s P(1,0,M) \tag{5}$$

$$\begin{aligned}[(M-i-j)(\lambda_1+\lambda_2)+\mu_1+\lambda_s]P_1(0,i,j) &= (M-i-j+1)\lambda_1 P_1(0,i-1,j) \\ &+ (M-i-j+1)\lambda_2 P_1(0,i,j-1) \\ &+ \mu_1 P_1(0,i+1,j)+\mu_2 P_2(0,i,j+1) \\ &+ \mu_s P(1,i,j) \qquad i=1,2,\ldots,M-1 \\ &\qquad\qquad j=1,2,\ldots,M-1 \\ &\qquad\qquad i+j\le M\end{aligned} \tag{6}$$

$$[(M-i-j)(\lambda_1+\lambda_2)+\mu_2+\lambda_s]P_2(0,i,j) = (M-i-j+1)\lambda_1 P_2(0,i-1,j)$$
$$+ (M-i-j+1)\lambda_2 P_2(0,i,j-1) \qquad i=1,2,\ldots,M \quad j=1,2,\ldots,M \quad i+j\leq M \tag{7}$$

$$[M(\lambda_1+\lambda_2)+\mu_s]P(1,0,0) = \lambda_s P(0,0,0) \tag{8}$$

$$[(M-i)(\lambda_1+\lambda_2)+\mu_s]P(1,i,0) = \lambda_s P_1(0,i,0)$$
$$+ (M-i+1)\lambda_1 P(1,i-1,0) \qquad i=1,2,\ldots,M \tag{9}$$

$$[(M-j)(\lambda_1+\lambda_2)+\mu_s]P(1,0,j) = \lambda_s P_2(0,0,j)$$
$$+ (M-j+1)\lambda_2 P(1,0,j-1) \qquad j=1,2,\ldots,M \tag{10}$$

$$[(M-i-j)(\lambda_1+\lambda_2)+\mu_s]P(1,i,j) = \lambda_s P_1(0,i,j)+\lambda_s P_2(0,i,j)$$
$$+ (M-i-j-1)\lambda_1 P(1,i-j,j)$$
$$+ (M-i-j+1)\lambda_2 P(1,i,j-1) \qquad i=1,2,\ldots,M \quad j=1,2,\ldots,M \quad i+j\leq M \tag{11}$$

The number of equations required to describe the above system is $M(M+1)+1+((M+1)(M+2))/2$. The above equations can be written as

$$\underline{AP} = \underline{0} \tag{12}$$

where $\underline{A}$ is a square matrix of dimension $M(M+1)+1+((M+1)(M+2))/2$ and its elements are the probability coefficients of Equations (1)-(11). $\underline{P}$ is a column vector which represents the steady-state probabilities and $\underline{0}$ is a null vector whose elements are zeros.

The steady-state solution of (12) is obtained by imposing the boundary condition (13).

$$P(0,0,0 + \sum_{j=0}^{M} \sum_{i=0}^{M-j} P_1(0,i,j) + \sum_{j=0}^{M} \sum_{i=0}^{M-j} P_2(0,i,j) + \sum_{j=0}^{M} \sum_{i=0}^{M-j} P(1,i,j) = 1 \tag{13}$$

As indicated in the literature, closed form solutions for the measures of effectiveness of the system such as machine availability and robot utilization do not exist. Therefore, numerical solutions of these equations are obtained for different values of the parameters, M, λ, and μ. The machine availability (M.A.) and robot efficiency (R.E.) are defined as:

$$M.A. = 1 - \{E[N_1] + E[N_2]\}/M \tag{14}$$

$$R.E. = 1 - P(0,0,0) \tag{15}$$

where

$$E[N_1] = \sum_{j=0}^{M} \sum_{i=0}^{M} i(P_1(0,i,j) + P_2(0,i,j) + P(1,i,j)) \qquad i+j \le M \tag{16}$$

$$E[N_2] = \sum_{j=0}^{M} \sum_{i=0}^{M} j(P_1(0,i,j) + P_2(0,i,j) + P(1,i,j)) \qquad i+j \le M \tag{17}$$

The total cost rate per machine for repairs and downtime is

$$C(M) = R/M + F\{E[N_1] + E[N_2]\}/M \tag{18}$$

From equation (8)

$$P_2(1,0,0) = \frac{\lambda_s}{M(\lambda_1+\lambda_2) + \mu_s} \tag{19}$$

The optimal number of machines (N^*) to allocate to a robot to minimize $C(N)$ is determined by satisfying inequality (20)

$$C(N^*-1) > C(N^*) < C(N^*+1) \tag{20}$$

2. POLICY B:

Under this policy, no priority is assigned to either types of failures (or service calls). The steady-state equations which describe this policy are:

$$[M(\lambda_1+\lambda_2)+\lambda_s]P(0,0,0) = \mu_2 P(0,0,1)+\mu_1 P(0,1,0)+\mu_s P(1,0,0) \tag{21}$$

$$\begin{aligned}[(M-i)(\lambda_1+\lambda_2)+\mu_1+\lambda_s]P(0,i,0) = {} & (M-i+1)\lambda_1 P(0,i-1,0) \\ & + \mu_s P(1,i,0)+\mu_2 P(0,i,1) \\ & + \mu_1 P(0,i+1,0) \qquad i=1,2,\ldots,M-1\end{aligned} \tag{22}$$

$$(\mu_1+\lambda_s)P(0,M,0) = \lambda_1 P(0,M-1,0)+\mu_s P(1,M,0) \tag{23}$$

$$\begin{aligned}[(M-j)(\lambda_1+\lambda_2)+\lambda_s+\mu_2]P(0,0,j) = {} & (M-j+1)\lambda_2 P(0,0,j-1) \\ & + \mu_1 P(0,1,j) + \mu_2 P(0,0,j+1) \\ & + \mu_s P(1,0,j) \qquad j=1,2,\ldots,M-1\end{aligned} \tag{24}$$

$$(\mu_2+\lambda_s)P(0,0,M) = \lambda_2 P(0,0,M-1) + \mu_s P(1,0,M) \tag{25}$$

$$\begin{aligned}[(M-i-j)(\lambda_1+\lambda_2)+\mu_1+\mu_2+\lambda_s]P(0,i,j) = {} & (M-i-j+1)\lambda_1 P(0,i-1,j) \\ & + (M-i-j+1)\lambda_2 P(0,i,j-1) \\ & + \mu_1 P(0,i+1,j) + \mu_2 P(0,i,j+1) \\ & + \mu_s P(1,i,j) \qquad i=1,2,\ldots,M \;,\; j=1,2,\ldots,M \;\; i+j \le M\end{aligned} \tag{26}$$

$$[M(\lambda_1+\lambda_2)+\mu_s]P(1,0,0) = \lambda_s P(0,0,0) \tag{27}$$

$$[(M-i)(\lambda_1+\lambda_2)+\mu_s]P(1,i,0) = \lambda_s P(0,i,0) + (M-i+1)\lambda_1 P(1,i-1,0) \qquad i=1,2,...M \tag{28}$$

$$[(M-j)(\lambda_1+\lambda_2)+\mu_s]P(1,0,j) = \lambda_s P(0,0,j) + (M-j+1)\lambda_2 P(1,0,j-1) \qquad j=1,2,...M \tag{29}$$

$$[(M-i-j)(\lambda_1+\lambda_2)+\mu_s]P(1,i,j) = \lambda_s P(0,i,j)+(M-i-j+1)\lambda_1 P(1,i-1,j) + (M-i-j+1)\lambda_2 P(1,i,j-1) \qquad i=1,2,...,M;\ j=1,2,...,M;\ i+j\le M \tag{30}$$

The number of equations required to describe the above system is $2(2M+1+M(M-1)/2)$ which can be written as given in (12). The difference lies in the dimension of the matrix A which becomes $2(2M+1+M(M-1)/2)$. The steady-state solution of equations (21)-(30) is obtained by imposing the following boundary condition.

$$P(0,0,0) + \sum_{j=0}^{M} \sum_{i=0}^{M-j} P(0,i,j) + \sum_{j=0}^{M} \sum_{i=0}^{M-j} P(1,i,j) = 1 \tag{31}$$

Again, we obtain numerical solutions for different values of the parameters M, λ, and μ. The machine availability (M.A.) and robot efficiency (R.E.) are defined as follows:

$$M.A. = 1 - E[N]/M$$

where

$$E[N] = \sum_{k=0}^{1} \sum_{j=0}^{M} \sum_{i=0}^{M} (i+j)P(k,i,j) \tag{32}$$

$$R.E. = 1 - P(0,0,0) \tag{33}$$

The optimal number of machines (N*) to be allocated to a robot such that the total cost is minimized is obtained by equation (20).

RESULTS AND CONCLUSIONS

From Figures 3 and 4 it is shown that the optimum number of machines assigned to the robot is a function of ρ_1, ρ_2, ρ_s and the priority of repair (service) of each type of failure. For example, in manufacturing cells where $\rho_1 > \rho_2$ a fewer number of machines should be assigned to the robot than those where ρ_2 is greater than ρ_1.

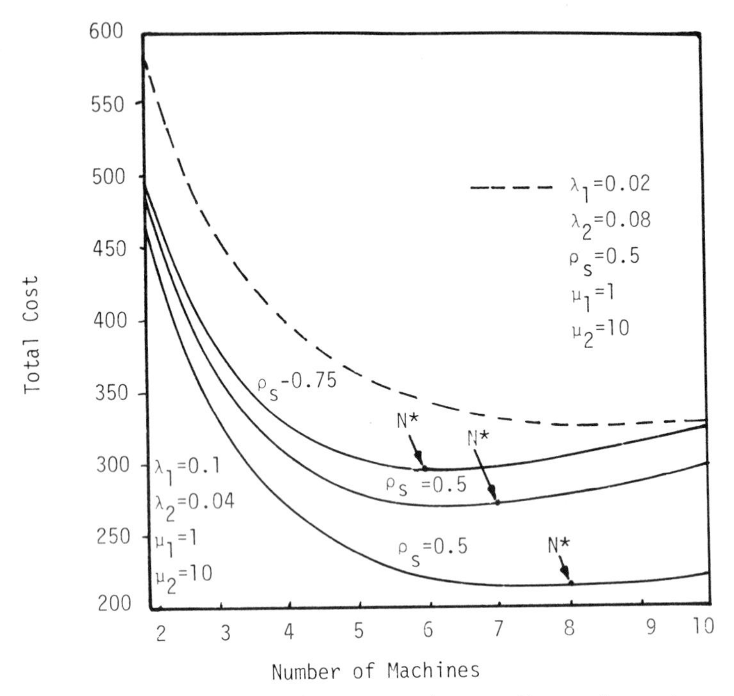

Fig. 3. Effect of ρ_1, ρ_2 and ρ_s on the assignment of the machine under Policy A.

In this paper, we have presented an approach for determining the optimal assignment of machines to a robot in manufacturing cells. The effect of the repair policy (priority opposite to no priority) of failures on the assignment has also been presented.

REFERENCES

1 E.A. Elsayed and J. Norton, Methodologies for Improving the Manufacturing of the Cross Deck Pendant, Working Paper No. 84-117, Department of Industrial Engineering, Rutgers University, 1984.

2 H. Schmigalla and W. Rodeck, Simulation of Manufacturing Cells with Industrial Robots, Proceedings of the VII International Conference on Production Research, Institute of Industrial Systems Engineering, University of Novi Sad, Yugoslavia, 1981, pp. 657-661.

3 K.E. Stecke, Machine Interference: Assignment of Machines to Operator, Handbook of Industrial Engineering, Edited by H. Salvendy, John Wiley & Sons, N.Y., 1982, pp. 3.5.1-3.5.43.

4 K.E. Stecke and J.E. Aronson, Review of Operator/Machine Interference Models, Working Paper No. 355, Graduate School of Business Administration, University of Michigan, 1983.

5 W.R. Wright, Machine Interference, Part I, Mechanical Engineering, Vol. 58, No. 8, 1936, 510-514.

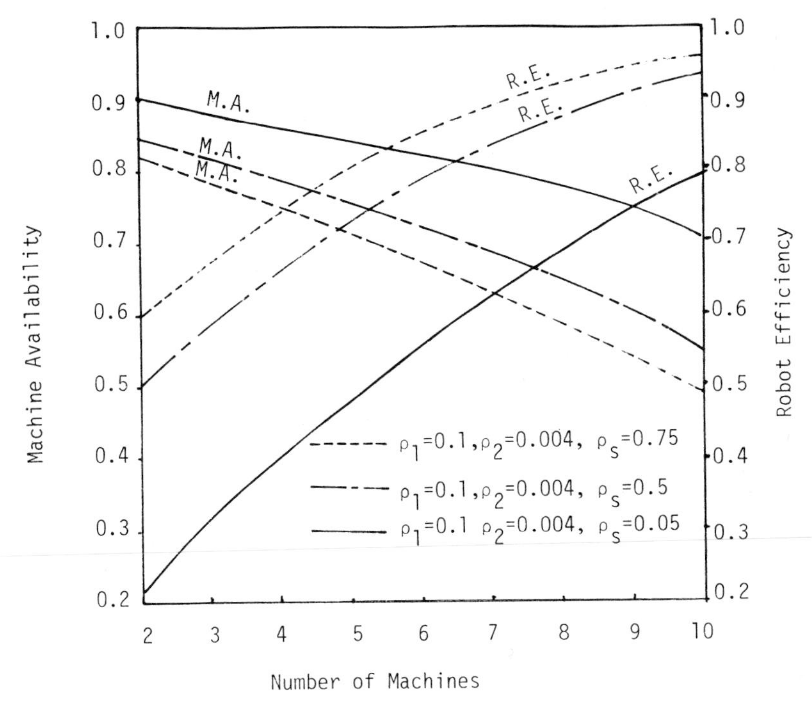

Fig. 4. M.A. and R.E. vs. the number of machines assigned to the robot under Policy A.

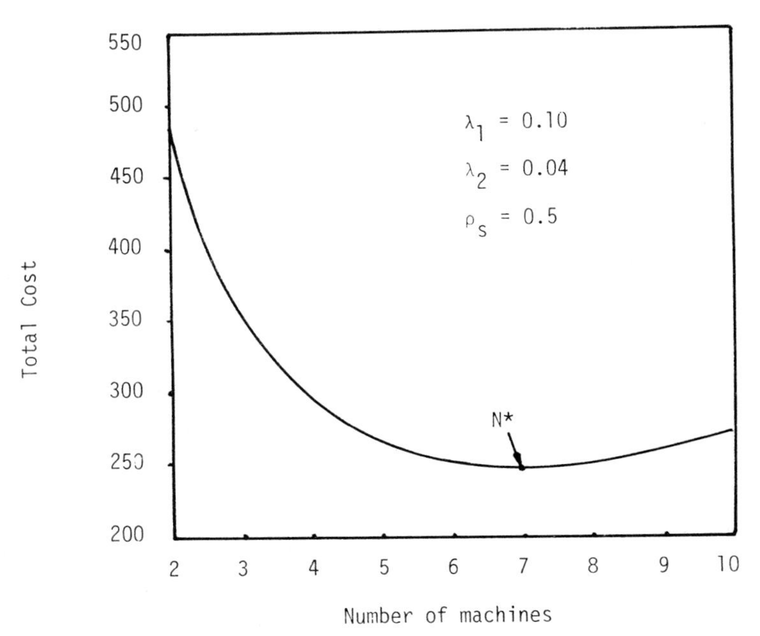

Fig. 5. Optimum assignment of machines under Policy B.

6 J.G. Miller and W.L. Berry, Heuristic Methods for Assigning Men to Machines: An Experimental Analysis, AIIE Transactions, Vol. 6, No. 2, 1974, pp. 97-104.

7 R. Dube and E.A. Elsayed, A Multi-Machine Labor Assignment for Variable Operator Service Times, Computers and Operations Research, Vol. 6, No. 3, 1979, pp. 147-154.

8 D.C. Palm, The Assignment of Workers in Servicing Automatic Machines, Journal of Industrial Engineering, Vol. 9, No. 1, 1958, pp. 28-42.

9 H. Ashcroft, The Productivity of Several Machines Under the Care of One Operator, Journal of Royal Statistical Society B, Vol. 12, No. 1, 1950, pp. 145-151.

10 F. Benson and D.R. Cox, The Productivity of Machines Requiring Attention at Random Intervals, Journal of the Royal Statistical Society B, Vol. 13, 1951, pp. 65-82.

11 N.K. Jaiswal and K. Thiruvengadam, Simple Machine Interference with Two Types of Failures, Operations Research, Vol. 11, 1963, pp. 624-636.

12 V. Hodgson and T.L. Hebble, Nonpreemptive Priorities in Machine Interference, Operations Research, Vol. 15, No. 2, 1967, pp. 245-254.

13 G.H. Reynolds, An M/M/m/n Queue for the Shortest Distance Priority Machine Interference Problem, Operations Research, Vol. 23, No. 2, 1975, 325-341.

14 E.A. Elsayed, An Optimum Repair Policy for the Machine Interference Problem, Journal of Operational Research Society, Vol. 32, No. 9, 1981, pp. 793-801.

15 R.A. Reich, A Simulation Approach to the Solution of a Machine Interference Problem, M.S. Thesis, Pennsylvania State University, 1964.

16 G.E. Haagensen, The Determination of Machine Interference Time Through Simulation, The Western Electric Engineer, Vol. 14, No. 2, 1970, pp. 35-40.

17 D.R. Freeman, S.V. Hoover, and J. Satia, Solving Machine Interference by Simulation, Journal of Industrial Engineering, Vol. 5, No. 7, 1973, pp. 32-38.

18 A. Garcia-Diaz, G.L. Hogg, and F.G. Tari, Combining Simulation and Optimization to Solve the Multimachine Interference Problem, Simulation, 1981, 193-201.

19 R.D. Guild and R.J. Hartnett, A Simulation Study of a Machine Interference Study of a Machine Interference Problem with M Machines Under the Care of N Operators Having Two Types of Stoppage, International Journal of Production Research, Vol. 20, No. 1, 47-55.

20 D.J. Medeiros and R.P. Sadowski, Simulation of Robotic Manufacturing Cells: A Modular Approach, Simulation, 1983, pp. 3-12.

21 Karmeshu and N.K. Jaiswal, A Machine Interference Model with Threshold Effect, Journal of Applied Probability, Vol. 18, 1981, pp. 491-498.

22 E.A. Elsayed and B. Dhillon, Time-Dependent Solutions of the Machine Interference Problem, Modeling and Simulation, Vol. 13, Part 4, 1982, pp. 1517-1521.

STUDIES TO ADVANCE THE APPLICATION OF ROBOTIC WELDING

Middle, J E, and Sury, R J

Department of Engineering Production, Loughborough University of Technology, United Kingdom

ABSTRACT

Investigations have been directed at appraising the technological and economic potential for adopting robotic electric arc welding in a number of companies, as an extension of the Department of Engineering Production, Loughborough University of Technology, Teaching Company Programmes. By reference to industrial feasibility studies, the paper discusses problems encountered in application to small batch fabrication companies and suggests means of alleviating these difficulties. Results of these studies encourage the wider use of robot arc welding in this important sector of the fabrication industry.

INTRODUCTION

By far the largest proportion of work undertaken in the fabrication industry is of a small batch nature and this offers significant potential for the application of robotic arc welding. Larger volume requirements naturally assist the case for investment in robotic facilities. In either situation the decision to adopt robotic welding hinges principally on economic, technological and organisational considerations. It is on these factors that one of the Department of Engineering Production Teaching Company programmes at Loughborough University of Technology has focussed. The aim has been not only to appraise the case for robotic arc welding but to assist its adoption.

Funding from the Science and Engineering Research Council and the Department of Trade and Industry has enabled study and development to proceed in collaboration with a number of companies. In the main the firms have been concerned with small batch fabrication; a few companies have dealt with more substantial volume and some have sought the advantage in manufacturing flexibility which planning based on small batches can provide. This paper draws on work undertaken in the programme and seeks to spotlight points which can assist wider application.

BASIC REQUIREMENT AND SOURCES OF ERROR

The basic requirement for successful automatic welding, and indeed manual welding, is that the arc be positioned, orientated and manipulated correctly relative to the joint. This must be done within the tolerance of the welding process being employed. It must also have regard to the quality standards being applied.

The sources of variability in robotic arc welding are:

(a) Accuracy of component parts manufacture and assembly. Thus attention must be given to parts manufacturing processes, assembly tooling and quality control to ensure that dimensional tolerances on assemblies for robot welding can be maintained within acceptable limits.

(b) Location of assemblies on the work manipulator. Additional holding fixtures and a precise fixture - manipulator location system are required in the manner of those commonly used in NC machining systems.

(c) Robot and manipulator errors in the repeatability of taught points. An acceptable level of repeatability is often the major system specification requirement for robot arc welding facilities; a typical requirement is in the order of $\pm$ 0.5 mm. However, the process technology occasionally permits wider limits as, for example, with Ransome, Sims and Jeffrys welding of plough frogs. A robot normally associated with spray painting and having a repeatability of $\pm$ 1.5 mm has been used successfully for this welding application.

(d) Accuracy of robot pendant teaching. Some degree of human error inevitably arises in the visual location arc position relative to joint location.

(e) Effects of distortion during welding. As with manual welding, careful attention is required to be given to weld sequencing. Due to the consistency of distortion occurring in automatic welding, this can largely be "programmed out" (by editing the end point of the last weld run and start point of the next) as the first component is welded one step at a time. Such a procedure has been demonstrated to be very effective in reducing the effect of distortion when robot welding even large and complex assemblies.

These sources of error are common to all applications of robotic arc welding. However, the development cost incurred, in limiting their cumulative effect to acceptable levels in a small batch fabrication situation, can be considerable in view of the many different assemblies that may be involved.

ECONOMIC ASPECTS

Alongside investment costs the economic justification for robotic arc welding is related particularly to the amount of welding a company carries out (that is, annual weld metal deposition or weld length), the time required to perform this and the relative efficiency with which the welding can be done by robot compared with a manual welder. A robot can be expected to be welding (arc-on) for up to, say, 80% of the overall cycle time (arc duty cycle) whereas manual welders typically achieve only about 20%. It has been found that few companies quantify this important productivity parameter.

Various practical welding trials have been carried out in the Department of Engineering Production using a Cincinnati Milacron T_3 robot system. Fabrications from collaborating companies have ranged over bicycle frames, crane gear boxes, manhole covers, lawn mower cutting blades, baking ovens, explosion proof chambers, coal burners, quarrying equipment, beer barrels, conveyors. Fig 1 illustrates welding trials on an excavator shovel.

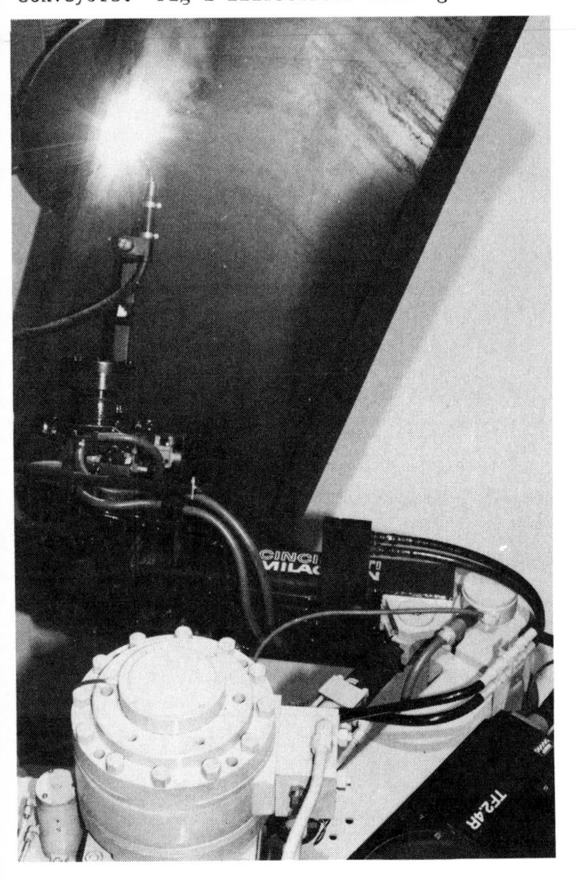

Fig. 1. Making a trial weld during teaching for robot welding of excavator shovel.

Examples of reduction in welding operation cycle times are given in Table 1.

Arc duty cycle measures were derived for robotic welding. These are compared in Table 2 with the in-plant manual welding measures. From this sample of assemblies the average improvement in arc duty cycle using robotic welding was in the order of 3.5 to 1.

TABLE 1 Examples of Reduction in Welding Operation Cycle Times

Component	Robot		Manual	
	Cycle Time	Arc Duty Cycle	Cycle Time	Arc Duty Cycle
	mins	%	mins	%
Hoist Anchor Beam	4.15	72	13	23
Lawn Mower Blade Cylinders (2 operations)	1.87	60	4.9	19
Conveyor Support Leg	6.92	79	22	23
Excavator Shovel	58	82	186	20
Fire Proof Chamber	134	78	230	45
Gearbox Case	37	79	178	17

TABLE 2 Manual and Robotic Arc Duty Cycles at 9 Small Batch Fabrication Companies

Company	Number of Components Assessed	Average Arc Duty Cycle	
		Manual	Robot
I	3	34	85
II	2	22	82
III	1	17	91
IV	2	26	65
V	2	14	68
VI	4	24	77
VII	1	21	58
VIII	1	19	73
IX	2	23	86

Scope for further improvement exists in several cases through robot programme refinement or through using an alternative robot. Investment, for example, in a second manipulator would improve the arc duty cycle through permitting the removal of an assembly and the mounting of the next set of components while welding is proceeding. Overall the studies have indicated a potential for a significant reduction in cycle time with marked opportunities for greatly improved work flow.

Fig 2 shows a simplified form of break-even analysis for the introduction of robot welding at a sub-contract fabricator when applied to a single sub-assembly, an excavator shovel, which is manufactured in large volume on a more or less continuous basis. Technological viability of this application requires some development to improve weld process tolerance through changes in joint design and weld procedure, tighter Quality Control in the manufacture of some parts and improved assembly fixture design and their maintenance. These pre-production and in-production development costs, together with programming cost are, for the most part, once and for all and spread over the system/product life. The organisational and development costs have not been so severe as to influence the company decision to implement robot welding.

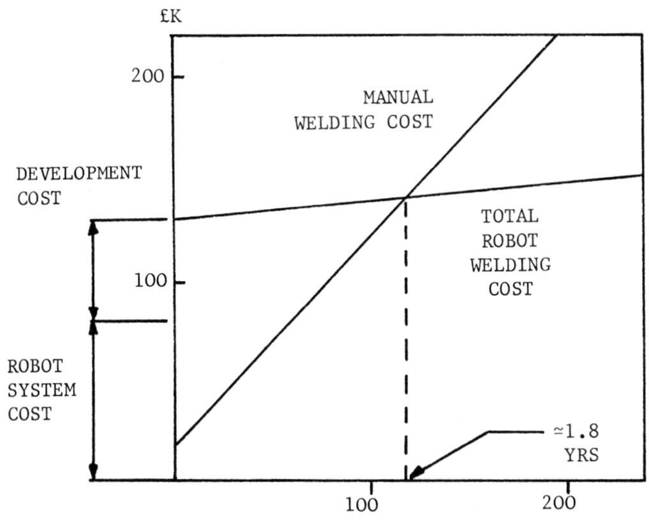

WELD LENGTH (x 1000 METRES)

WELD LENGTH	18m PER PART
ANNUAL OUTPUT	3500 UNITS
MANUAL CYCLE	186 mins @ 20% ARC DUTY CYCLE
ROBOT CYCLE	58 mins @ 82% ARC DUTY CYCLE
ESTIMATED COST SAVING	£22 PER PART
ANNUAL ROBOT USAGE	3400 HRS

Fig. 2. Breakeven Analysis - Excavator Shovel Example.

Cost recovery is more protracted however with multiple product fabrication in small batches. This is illustrated in Fig 3 showing a similar analysis of break-even cost for the application of robotic welding for some 80 different assemblies in a particular company. At first sight the total workload generated from these sub-assemblies looked likely to make a robotic arc welding system economically attractive; an average manual arc duty cycle of less than 20% provides further strong support.

In this case however similar attention to the control of cumulative errors in the joint/arc location system was required over the much larger number of parts. The development costs were such that the total implementation costs were almost twice the cost of the robotic welding facilities. The payback period was found to be double that of the single assembly example.

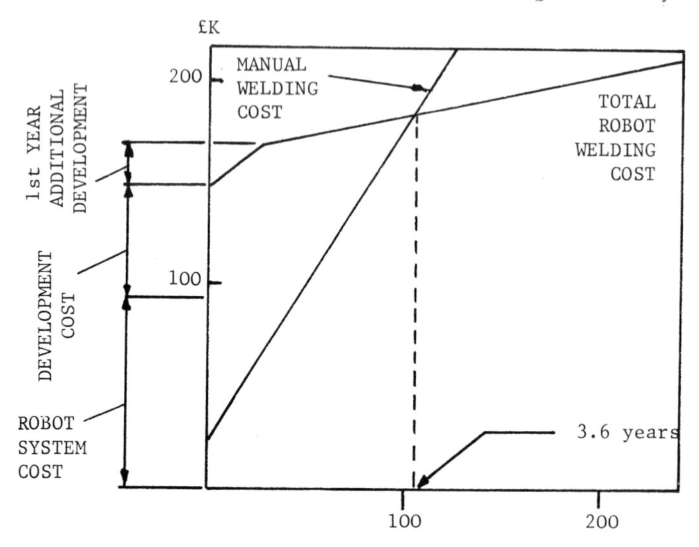

WELD LENGTH	28000m PER ANNUM
ANNUAL OUTPUT	APPROX 8100 UNITS
AVERAGE MANUAL ARC DUTY CYCLE	17%
AVERAGE ROBOT ARC DUTY CYCLE	83%
ESTIMATED COST SAVINGS	£45000 PER ANNUM
ANNUAL ROBOT USAGE	1800 HRS

Fig. 3. Breakeven Analysis - Small Batch Example.

Two to two and a half times robot cost has been found to be the normal order of total implementation cost for the several small batch fabricators for whom feasibility studies have been conducted. Such high development costs militate heavily against economic justification and arise largely from the need to refurbish or replace existing assembly jigs and from the large number of new holding fixtures needed. Additionally, in most cases, some attention to parts manufacturing practice, technology and design is needed. Recommendations for changes to parts manufacture have included implementation of CNC thermal cutting to replace optical systems and the introduction of CNC turret punching facilities. Such technology has been shown to be cost effective in its own right in addition to improving dimensional variabilitity of assemblies for robot welding.

Introduction of improvements in manufacturing to facilitate robot welding will often result in benefits elsewhere in the manufacturing system. More consistent fabrications offered to machining or assembly and reduced machining allowances are examples but it has not been possible to quantify these advantages comprehensively in the studies undertaken.

A further problem exists for the small batch fabricator. The robot system must be programmed for each of the jobs involved. In the case of the company for which the second break-even analysis has been illustrated, pendant teaching of the robot would take, on average, approximately two days per job; that is a total of 160 days for the 80 jobs identified as suitable for robot welding. This is not only a further heavy development cost burden but it also significantly reduces the productive availability of the robot during the introductory period.

Additionally a common feature encountered with small batch fabricators is that design changes occur frequently and significant numbers of new jobs are continually introduced. This has the advantage that, in the course of time, the majority of fabrications destined for production by robot welding will have been designed and production engineered to suit this technology. On the other hand a considerable amount of robot re-programming and programming of new jobs is required, representing a significant recurrent cost and marked reduction of the productive availability of the system. Fig 3 includes the influence of these additional recurrent costs.

SEAM TRACKING SYSTEMS

It is clear that research and development underway with seam tracking systems could lead to problems associated with excess errors between the arc path and joint line being largely eliminated. For some types of welding requirement the more recently available technology may well achieve this now - as, for example, the Oxford University/British Leyland/GEC structured light and vision based seam tracking and adaptive process control system. This has been demonstrated recently with sheet metal lap jointed assemblies.

However, the commercially available seam tracking systems all have limitations for the small batch fabricator represented in this study, with the wide variety of size, material thickness and complexity of their welded assemblies. These limitations include the use in some cases of a two pass technique, access difficulties for probes and sensory devices, the need to ensure that joint designs have suitable features for sensing, the present high purchase cost of some systems. Whilst developments in the area of arc welding sensory systems for robot applications are encouraging, wholly suitable systems for the breadth of applications required are nevertheless sufficiently distant to justify encouragement of industrial implementation of current generation sensorless robot arc welding systems.

Hence the studies referred to in this paper aimed at providing fabrication companies with a better understanding of the demands of robot welding and the means by which successful adoption can be achieved. Some points have already been touched on but can be amplified in regard to weld tolerance, manufacturing capability studies and design.

WELD TOLERANCE

It would appear that comparatively few fabrication companies have given attention to obtaining a quantified understanding of the tolerance of their weld processing to variation in arc location, plate edge misalignment or gaps in fit-up. This is surprising since this can have a marked effect on quality and cost even of manual welding. Such knowledge can be crucial to the economic use of automatic welding insofar as it effectively establishes the requirements for accuracy in the manufacture of parts and assembly and in the robot system specification.

The fundamental requirement here is that the combined variability arising from the manufacture of parts, the assembly of parts, the accuracy and repeatability of the robot system must not exceed the weld process tolerance.

Fabrication companies typically assume a "rule of thumb" tolerance of $\pm$ one electrode wire diameter. It has been demonstrated that this can over or under-estimate actual process tolerance. A necessary part of feasibility studies carried out in companies has been to establish process tolerance for each weld procedure to be used in robot welding relative to the quality standards applied. Time and development cost limitations prohibit exhaustive testing of all combinations of variables but sufficient work can be done to give useful data. Figs 4a and 4b show typical data produced, giving envelopes of acceptable combinations of assembly and fit-up variables. A "catalogue" of such process data is being established as various feasibility studies are undertaken. It is intended to publish this work when more complete coverage of possible Gas Metal Arc Welding variables has been established.

It should also be emphasised that robot welding will often be possible under conditions beyond physiological limits of a manual welder. Thus not only can robotic welding improve productivity through increased deposition rate but may also provide larger weld process tolerance.

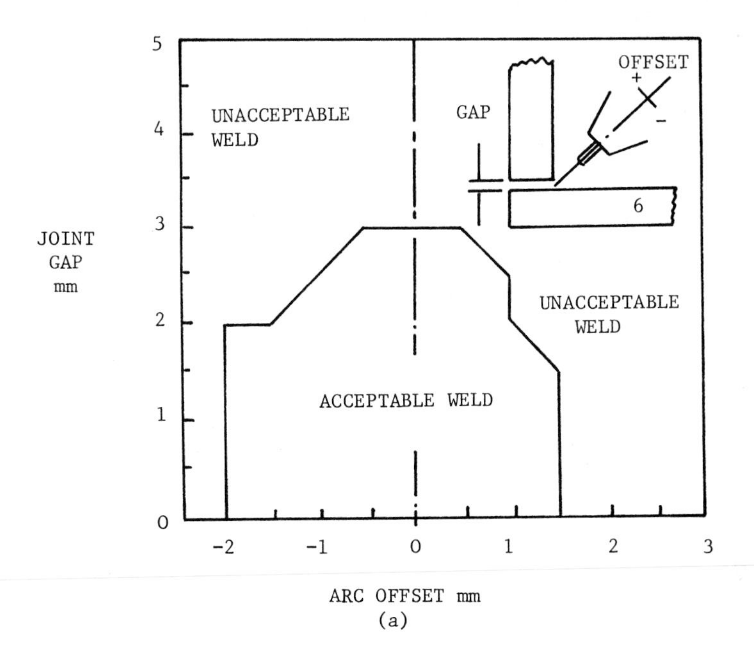

(a)

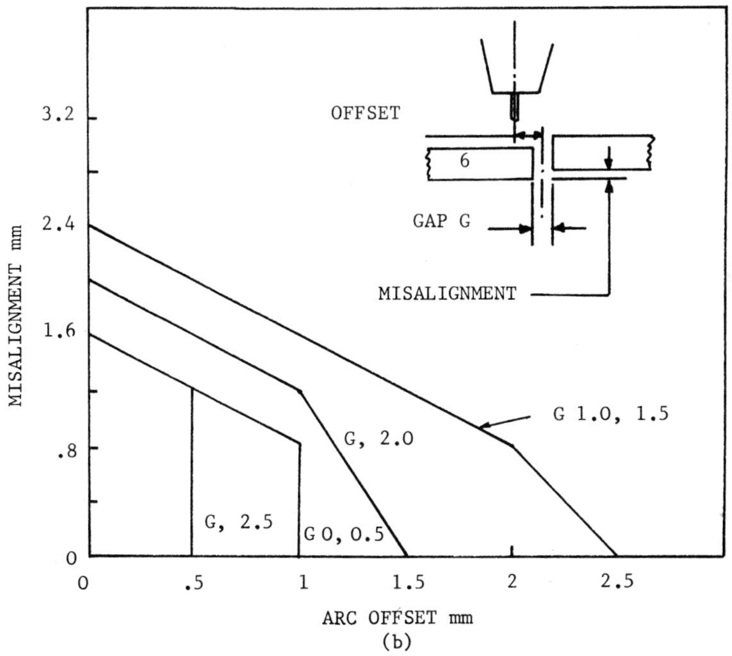

(b)

CONDITIONS: 1.6mm Dia. METAL CORED WIRE. 330 Amps. 25V, 12.5mm SEC^{-1}.
Shield - AR + 20% Co_2

Fig. 4. Influence of fit-up and arc position on weld process tolerance (a) fillet welds (b) butt welds.

MANUFACTURING CAPABILITY STUDIES

When weld process tolerance is more precisely known and cumulative robot system errors are defined then permissible parts and assembly manufacturing variability can also be defined. Sampling of critical dimensions of parts and assemblies, taken from oxy-gas cutting, shearing and forming processes and from assembly tooling, in each of the companies involved in our collaborative studies has enabled assessment of their capabilities of matching the requirements of robot welding.

The analyses have indicated that some companies could expect 95% of weld joints to be located within the process tolerance with a high level of confidence. In these cases little development work has been necessary; manual welding has been used for the few remaining joints. Other cases have been less satisfactory. The analyses have shown where attention has been needed to effect improvement. Major problem areas are assembly fixturing, oxy-gas burning and formed components.

Assembly fixturing for manual welding of assemblies, in which close dimensional control is often not critical to function, is frequently designed simply to close up the joints. In many of the examples studied some components were crudely located by eye. Different fixture design philosophy is required for robot welding application in that all component locations are critical. Robotic welding fixtures must therefore ensure that defined accuracies are achieved in the positioning of parts and in a precise location of fixture on the robot manipulator.

Close dimensional control of oxy-gas cut components requires careful attention to process parameters and cutting sequence. CNC systems readily facilitate this, which makes such systems technologically desirable for use in conjunction with automatic welding. It has been found that the higher cutting speeds, and other advantages, available from CNC systems provide cost justification of these machines, in addition to assisting in the implementation of robot welding. In particular, CNC plasma arc cutting can be an attractive proposition.

Dimensional accuracy of formed components n the studies undertaken has invariably been insufficient for robot welding. NC control of bending would show to advantage here but it has not been found possible to justify this on the strength of robot welding alone. The cost effective solutions have been either to design out formed components, or to leave welding of these components to a manual welder. In the case of the excavator shovel discussed earlier, welds in four locations were left to manual welding, 97% being done by robot.

DESIGN CHANGE

Changes to fabrication design have been introduced in several cases providing improved weld process tolerance, better robot access to joints and simpler and more accurate assembly.
Simple design changes as in Fig 5 were introduced at minimal cost. In this example problems of loss of control of weld penetration, due to variable gaps at the root of external corner fillet welds, were eliminated by simply increasing the size of some members with fixture modified to ensure a minimum overlap in all circumstances.

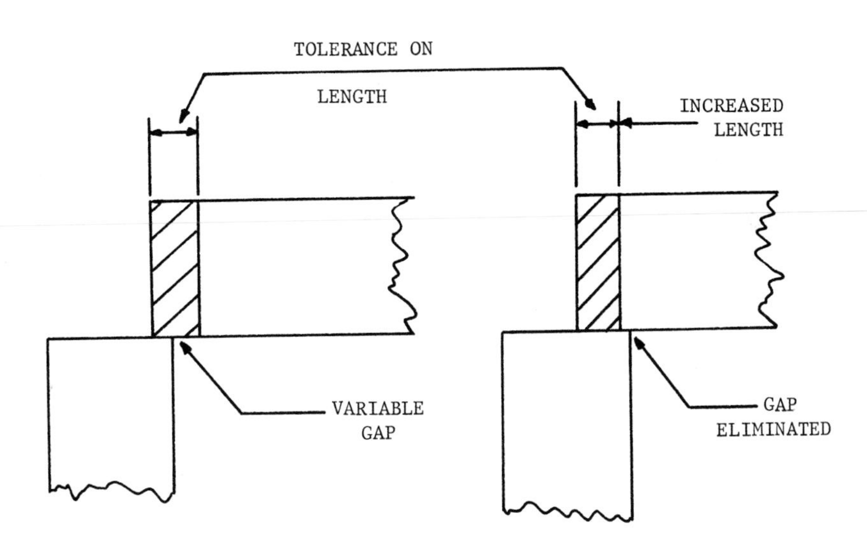

Fig. 5. Simple design change.

Fig 6 indicates a more radical design change introduced largely to make robot welding feasible. In this case the company manufactures a variety of processing equipment that includes welded chambers, such as ovens, in a number of types and sizes. Fig 6(a) shows the original construction which required considerable amounts of welding on the inside of the chamber, with difficult access by robot. It was also difficult to accurately locate the internal members. The new design, Fig 6(b), uses a simple self jigging principle for most of the members in the assembly, with all welds produced on the outside. Precise location of joints is assured through the use of a CNC combination turret punch and plasma cutting machine, and CNC shearing, for the manufacture of all parts. In this case formed components do not influence joint location. Introduction of the CNC plate processing facilities had been justified through reduced cost of parts manufacture and was not dependent on the use of robot welding.

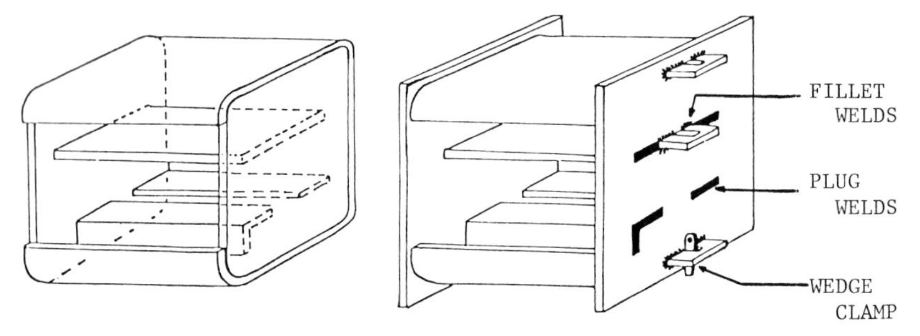

Fig. 6. Radical design change for CNC plate processing and robot welding. Assembly is self-jigging, all welding is external.

OFF-LINE PROGRAMMING

The problems of excessive time for pendant teaching of the robot referred to earlier can be alleviated through the use of off-line programming. Some of the companies involved in the programme are well advanced in the application of CAD/CAM systems and it is natural that they should wish to exploit the potential for off-line programming in conjunction with robot application. Two methods have been explored and are currently being further developed.

Interactive graphics is seen as the most appropriate method of off-line programming of welding; solid modelling and orthogonal viewing provide easy visualization of weld torch orientation. Off-line programming procedures based on the Graphical Robot Application Simulation Package (GRASP) developed at the Department of Production Engineering, Nottingham University have been employed. For this collaborative research, the Cincinnati Milacron T_3/Teledyne positioner system at Loughborough University has been modelled. Individual components and their holding fixtures can be added to the model in appropriate orientation, and the program produced by targeting the cursor onto the end points of weld runs and other appropriate locations to complete the robot path. Detection of possible collisions is assisted by the orthogonal viewing and computer routines.

Modelling of a gearbox casing, Fig 7, took six hours, and complete programming of the robot path approximately three days, similar to the time required for pendant teaching. Modelling and programming of a number of different jobs shows that, in the current state of development and with experience, off-line programming should take a similar time to on-line teaching. The system currently produces listings which must be manually entered to the robot controller in the manner of editing. Accuracy of programs produced is currently $\pm$ 3 mm, due primarily to inaccuracy in robot motion (as opposed to specified repeatability). Programs therefore require "fine-tuning" on-line. In the example shown this took approximately 4 hours, attention being needed only to points on the weld path.

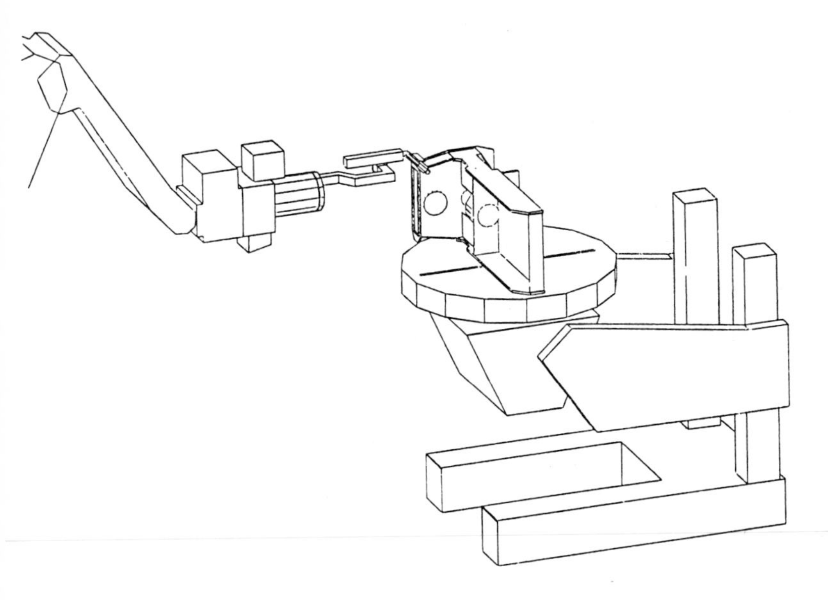

Fig. 7. GRASP model - crane gearbox.

Work is currently ongoing to establish the absolute accuracy of the robot system to improve the robot model, to establish direct communication to the robot controller and to provide estimation of cycle times and welding costs. The system can also be used to aid optimisation of workpiece orientation on the work positioner, fixture design and general workplace layout in order to provide minimum work and robot manipulation and best access for the robot.

A method has also been developed for application in companies that do not have access to powerful computing or CAD facilities. Vector analysis in the form of a matrix has been used to derive equations for formulating positional and orientational data for the end effector, the welding torch. Accuracy of program points produced is again $\pm$ 3 mm which is capable of improvement when compensation for robot inaccuracy can be introduced. The method is currently programmed in Basic to run on an Apple II computer, but can readily be modified to run on most modest micro-computers, and is easily inter-faced to the robot controller via the systems external computer function.

CONCLUDING REMARKS

Alongside wider industrial collaboration, in depth studies have been pursued in the sample of nine companies to which reference has been made in this article. One company has determined the nature of the robotic welding which will be incorporated in its flexible assembly system of lawnmower blades. Four companies producing fabrications in small batches now have robotic arc welding among proposals for future capital investment. Another company producing in larger volume are in an advanced stage of implementation of a two robot, two station system for welding various sizes of large assemblies of iron castings.

Implementation costs in these cases range from approximately £120K to £180K; payback ranges from less than two years up to about three and a half years. Robot teaching programming costs are reflected in these figures; off-line programming has not yet been implemented. Significant improvements in work flow rate and in reduced work in process will be achieved.

The studies indicate a potential for marked improvement in the methods engineering of the welding process. For the full potential of robotic arc welding to be realised, particularly for small batch fabrication companies, close attention must be given to weld tolerance, product design, parts manufacturing and assembly fixturing. Attention has been drawn to the significant advantage to be derived from arc duty cycle improvement; much higher levels of achievement can be obtained by the use of robotic welding.

The study overall has re-inforced the view that significant manufacturing engineering advantage and economic benefit can be obtained from the adoption of robotic welding for fabricating companies manufacturing under conditions of small batch production.

ACKNOWLEDGEMENTS

Grateful acknowledgement is made to SERC and DOI for the Teaching Company grant which enabled this work to proceed and to the management staff in the companies who so willingly collaborated in the study. Particular acknowledgement is made also of the significant contributions made by the two Teaching Company Associates, Russell Thorne and Phillip Wadworth.

APPLYING ROBOTICS IN A LIGHT INDUSTRIAL ENVIRONMENT

Doug Strong, Phd., P. Eng.
Bata Engineering, Batawa, Ontario

ABSTRACT

The application and usage of assembly robots in light manufacturing environments is of paramount interest today. As an important part of F.M.S. (Flexibility Manufacturing Systems) and C.I.M. (Computer Integrated Manufacturing), the successful implementation of assembly robot systems requires a careful study and analysis of present and proposed manufacturing methods. The results of such a study will provide both a better understanding of the present work methods and the requirements of specific tasks. In addition the results indicate which operations can be or are desirable to be accomplished by robots for reasons such as accuracy, safety, economy and consistency.

Specifications for a robot capable of performing the tasks required can be defined, along with whatever re-arrangement of work flow is necessary to implement a robot assembly system. With the assistance of a work study analyst from the shoe industry, an industrial analyst, and an engineering student, and using video-tape techniques, the author has completed an in depth study of assembly operations in a shoe factory.

The problems encountered are typical of mixed-batch, medium-volume levels of production. For this reason the methods for and the results of the study can be applied to many light manufacturing assembly operations.

General

This paper discusses the use of robots for assembly and machine loading functions in light industry. More specifically the paper discusses the analysis of the application of robots to production operations in the Shoe Industry.

To be cost effective the Robot and the jobs done must have several characteristics. One extremely important characteristic is that there must be enough jobs for the Robot to do to keep it fully occupied 12 to 16 hours per day for a couple of years. Another characteristic, the one discussed in this paper, is that the Robot must be easy to set up for short run production or batch processing operations, and it must be capable of handling variations in a product which go through mixed volume production lines.

No Robots were found which could be used effectively in shoe manufacturing. The available robots could not easily adjust for variations in part position, orientation or size which are problems typical of medium and mixed volume manufacturing. Also for most manufacturing operations the Robots moved too slowly to perform the work as quickly as a human operator.

The IBM assembly Robot with six degrees of freedom is the one Robot presently available which can adjust to its work situation. However, the ranges of part size and of part shape with which it can deal are too small for the Shoe Indusrty, while the Robot itself is too large and unweildy for that Industry.

SHOE PRODUCTION OPERATION ANALYSIS

A Robot System thought to be suitable for automating many production operations and using attainable technology was hypothesized. All the manufacturing operations of the Bata Footwear Company of Batawa, Ontario were then roughly analyzed to see which operations could be performed by the Robot. To aid in the analysis extensive video footage was taken of some of the operations. The definition of the Robot System and the analysis of the production jobs was of course an iterative procedure.

For the analysis a representative production day was chosen and all the operations for the main shoe styles going through the plant on that day were analyzed to obtain the number of operations which could be automated.

The results of the rough study are as follows .. Total production, 15,000 pairs per day in four major styles with the smallest group of a particular style, size and colour being ten pair. The total plant had about 1,000 employees of which office, supervisory, and sales totals 250, and direct production workers total 750. Of the production operations it was felt about 225 could be automated by the Robot. That is 30% of production or 22% of the total plant.

Even if the estimate was over optimistic by a factor of 2, 115 operations could still be performed by the hypothesized Robot.

THE HYPOTHESIZED ROBOT SYSTEM

The Robot System needs five major qualities to be useful in

the medium and mixed volume environment.

1. 7 Degrees of Freedom.

   The arm motion should move in all of X, Y and Z directions. The wrist should have pitch, yaw and roll actions, and the robot should be able to travel along a track.

   The Robot should also be light, compact, fast and accurate, although the accuracy need not be maintained while it is moving on the track. The robot should be able to move at the average speed of most workers when they perform the production operations.

2. Touch and Proximity Sensing.

   The proximity sensing could be any useful technique such as simple vision, electrostatic, air, electromagnetic, etc. The sensing would be used for safety, and to maintain relative accuracy.

   For the next several years, assembly robots would be most productive when they work together with human operators. This can only be achieved if the whole robot structure has proximity sensing and associated controls which stop the robot if it moves too close to any object or worker. Because the operation has to be set up very quickly in medium volume production, and because of the variations in product size in mixed volume production, it would be impractical to define in advance precisely where the robot should be under all conditions. The robot must be given approximate statements of where it should be and then it must determine for itself the exact position of the part that it is handling by using its touch and proximity sensing.

3. Interchangeable Tools and Grippers.

   In a medium volume production situation, especially a batch processing task, the robot is most usefully used if it can do several jobs in succession on a particular group of parts. Using this mode of operation, the robot spends more of its time actually performing job functions and less of its time finding and storing parts. To perform jobs in succession several tools may be required or a few different gripper structures may be required. The robot will be more productive if it can very quickly, easily and automatically change the tools and grippers.

In a mixed volume production situation, the range of size parts could be very large, and if this is the case it would be desireable to have tools or grippers of different sizes, which are otherwise identical, and the robot would automatically change to the appropriate size gripper or tool for the particular part going through the line.

4. Simple Mechano Set Style Jigs and Fixtures.
   A technique for creating jig shapes as required for the operations, along with a technique to add simple active parts which could move sections of the jig under machine control is required. It is also necessary to incorporate touch and proximity sensing in the jigs to give the robot more informaion about the production operation.
   The jig could be looked at as a second hand for the robot.
5. Off Line Programming.
   It should be possible to program the robot using an off line line programming technique which does not interfere with the production process.
   The programming system should also allow:-
   The interchangeability of robots at a work position.
   Operational error detection and recovery: techniques that allow the robot to adjust to and correct many problems occuring during the assembly process, such as replenished part stacks as shown in Fig. I.
   Compliance: Techniques allowing the robot to adjust to inaccurately positioned parts, and allowing the robot to slide along in accurately defined surfaces.

With all these qualities this "intelligent" Robot System has still very limited abilities in that its dexterity is extremely limited relative to a human worker, and even with the sensing given to it, the robot has very little information about its production environment. The robot could be roughly equivalent to a blind-folded person using tongs.

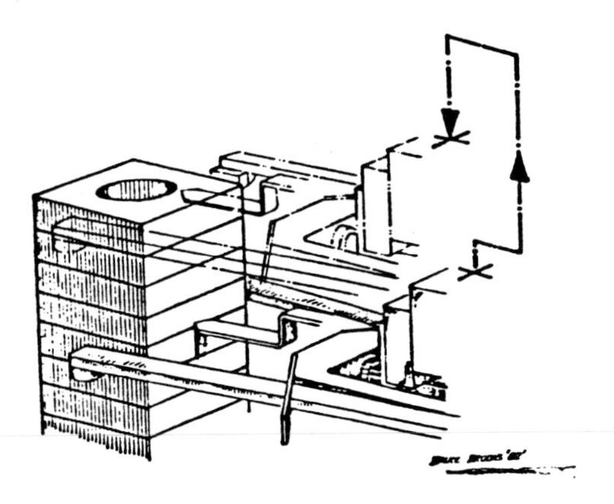

Fig. 1

DESCRIPTION OF THE SHOE PRODUCTION LINE

There are four major steps in shoe production: cutting and preparing parts for the shoe upper, sewing the shoe upper parts together, moulding and preparing the shoe soles, shaping the upper into the shoe shape and attaching the sole.

The last steps, shaping the upper and attaching the sole, are called the shoe production line. The operations performed on this line will be described briefly to give a feel for the range of jobs performed. Then two operations will be described in more detail and their robotization using the hypothetical robot system will be discussed.

The production line has about twenty different operations, all grouped around a conveyor which moves the shoe parts from one operation to the next. The process starts with the last or shoe shape, typically made of plastic. In a size range such as mens or womens, 13 different sizes of last are required. For shoes graded in width as well as length up to 65 different sizes of lasts are required to fully represent a shoe size range.

In the first operation an insole is tacked on the bottom of the shoe last for the style required. Next, a thermoplastic counter is inserted into the heel area of the upper, moulded and properly shaped to prepare the upper for placement on the last, using lasting machines the upper is stretched over the last and the bottom edge of the upper is wrapped around the edge of the insole and glued to it. This process is called lasting the upper and is performed in two steps, first at the toe and then the heel and waist.

The bottom of the upper is roughed, cemented and dried in preparation to accept the shoe sole. The glue on the upper and the shoe sole are heated to activate the glue, then the upper and sole are pressed together.

The shoe is removed from the last, and an additional soft insole or sock liner is added to the shoe. Laces are inserted. The shoe goes through a minor clean up operation and inspection to make sure that it has been produced properly, and then the shoe is packed.

You can see even from this very quick and sketchy description that there are a wide range of job types on the shoe production line, all well defined, but by machine standards the jobs are very variable and inprecise.

HEEL-WAIST LASTING OPERATION

The heel-waist lasting operation glues the bottom edge of the shoe upper to the insole at the heel and along the waist of the shoe. The previous operation has already attached the shoe upper over the last and has joined its lower edge to the insole at the toe.

The operator's task is to remove the shoes one at a time from the conveyor, place the lasts properly in the heel-waist lasting machine, activate the machine cycle, remove the shoe when the cycle is finished, and place the shoe back on the conveyor.

The operations are quite simple from a human point of view, but we should go over each step carefully to see how easily the robot system could be applied.

First picking up the shoe from the conveyor, the conveyor is composed of racks placed end to end, which are slowly dragged around the line by a moving chain. Each rack has 20 slots for holding shoes, 5 slots along the line by 4 slots high. Because of age and design, the position of each slot on the production line wanders and can be known only within + or - a few inches.

The shoes, because they are not a regular easily definable shape, do not sit accurately in the slot of the rack.

Considering the accuracy problems, if the rack system is overhauled and straightened out where possible, the very best absolute position accuracy that could be obtained would be about + or - an inch in each of the X, Y and Z directions. However, this inaccuracy is no problem to the human operator.

If a proximity detector system were able to measure the position of three corners of each rack, then the position of each slot could be known to within + or - a quarter of an inch. Then if the size of each shoe is known, the position of the toe of the shoe within the rung of the rack could be known within + or - half an inch. That means that if the robot is able to search within + or - an inch range for the shoe, the shoe will always be located.

Once the shoe has been picked up, it is moved to the heel-waist laster. The problem now is, with the one hand, how to pick the lasted upper off and place the unlasted upper on the machine.

A possible solution would be to redesign the lasting machine to have two shoe mounting points and each one in turn would move into the lasting operation area. This solution would be easier to apply than designing a 2 gripper hand for the robot, and the operation would run faster than resequencing the operation.

The last with the shoe on it must be loaded into the machine upside down with the heel of the last facing towards the machine and a hole in the top heel area of the last dropping on top of a pin in the lasting machine mount point. Since the last when in the rack, was not accurately positioned or positionable, its exact position and orientation relative to the gripper will not be known.

The hole and pin can be defined so that a mismatch + or - a quarter of an inch would not upset anything. Some sort of cone guiding device would have to be used, at least in a left/right direction because the top of the last is no wider than the width of the hole and the pin could come up beside the last. In the backwards/forwards directions the robot could use a search motion to find the pin.

When the shoe to be lasted is moved into the lasting area of the machine, the lasting operation should be automatic. However, in practice the operator always grabs hold of the glue guns which are moved by the machine along the edge of the insole. The operator does this because once in a while the glue guns jump out of their tracks and do not glue the shoe correctly. This problem has never been fixed by the machinery company because the operator is available to assist the machine. For operation with a robot, it is necessary to modify the automatic process so that either it never skips of that it skips only very rarely, and when it does it notifies the production supervisor.

Self checking procedures must be incorporated throughout the operation system to ensure that the shoe has been picked up properly, that its orientation is approximately correct, that it has been properly placed on the pin, that the gluing operation has gone successfully and the last upper returned to the correct slot in the conveyor rack. If any of these operations are prone to failure, that is more often than once every hundred times, it is necessary to have an operational error correction technique with the aim of having the system call the operator no more than once every several thousand operations.

You can see from this example that even with sophisticated robot systems the simple job of machine loading for mixed batch production is still very complicated for many industrial situations and especially those in the shoe industry which is for the most part labour intensive.

OINING THE SOLE TO THE LASTED UPPER

The shoe sole has been glued and is on the moving conveyor, lso the shoe upper has been lasted and the bottom of the shoe pper has been glued.

Preceeding the joining operation the shoe sole and lasted pper are separately picked up from the rack and put into an ctivator, an oven with a rotating rack. (Fig. V) The heat in he activator causes the glue to work like contact cement. nother shoe upper and shoe sole that have been in the activator or awhile are picked up and placed together very carefully by and. The combination is placed in a press to force the upper nd the sole tightly together. Then the completed shoe is moved ack to the conveyor rack.

The problems of moving the shoe parts from place to place re very similar to the problems of the previous example.

The automation could be simplified and improved if the luing technology and technique could be changed slightly. It ay be possible to activate the glue quickly with a flash heater lement. The element could be briefly moved between the upper nd the sole in the jig before the upper is placed against the ole.

Using the flash heater technique a drop off and pick up peration could be removed for each of the upper and the sole andling. Also, the total operation sequence could be organized o allow the robot to change grippers for upper handling and sole andling through each operation sequence without significantly lowing the pace of the operation. This would allow the omplexity of gripper and wrist holder to be reduced.

As before, the grippers or parts of the grippers would have o be varied automatically when there is a significant size hange in the shoes going through.

The new problem in this operation is the actual placing of he sole and upper together. The alignment of the two parts is bvious to a person who knows the shoe style, but there is no intrinsic way of deciding the relative positions of the two parts.

However a fairly simple active jig system could be used to solve the problem for many different types of soles. One possible jig configuration has a fixed upper guiding structure used for guiding the upper and a slow speed adjustable lower, guiding structure for guiding the shoe sole. These structures could touch the front and one side of the shoe parts.

Machine controlled clamps would also be used to hold the sole in place while the upper is placed on it. Sensors would be used to make sure that the sole is properly positioned before the upper is brought down and also to make sure that the upper is being brought into position correctly.

The robot would pick up the activated sole, place it in the flat area near the front and side constraints and then slid it into the constraint region until the side and toe of the sole touch. The touching would be noted by a change of force level in the wrist of the robot and by sensors in the jig. Clamps would then be applied to hold the sole in place.

The robot hand would pick up the activated, lasted upper and holding it above the sole, push the upper against the toe and side constraints of the upper structure. Then the quick flash heating. The robot would then lower the lasted upper down onto the sole. Sensors would of course be used to make sure that each operation was performed correctly.

CONCLUSION

An assembly robot suitable for many jobs in light industry can be defined today and should be producable in the next few years. However a lot of work still has to be done to make such a robot truly useful and cost effective.

Of particular consequence will be the development of robots with:-

- tactile sensing for recognition, orientation, safety and physical interaction,
- vision to provide orientation and recognition data,
- intelligence for usage of the tactile and vision information,
- universal or multiple grippers,
- and minimal spatial intrusion.

Additionally, with the standardization of assembly technique in production operations, and the further integration of computerized production control and robotic systems, more potential successful solutions will become available.

COMPUTER AIDED MACHINING PROCESSES DESIGNING SYSTEM

E.WATLY[1] and S.KNAP[2]

[1]Institute of Organization and Management,Warsaw Technical University,85 Narbutta Str.,02-524 Warsaw /Poland/

[2]Institute of Organization and Management,Warsaw Technical University,85 Narbutta Str.,02-524 Warsaw /Poland/

ABSTRACT

The main goal of system is to increase efficiency of machining processes design, as well to shorten the processing time evaluation. It is realized through making objective answers to the following questions:

-How to divide machining processes into machining operations?

-How to allocate machining operations to machine tools?

A designer creates only the conception of the machining process. All other operations are realized by the computer system, so the time of machining processes design is reducted.

The idea of the system is based on the following remarks:

-Each surface that occures in machine parts can be defined and classified.

-Each machine part can be univocally described by means of a finite set of surfaces.

-Each surface can be obtained through a finite quantity of machining operation sequences.

-Each sequence can be performed by a finite quantity of machine tools.

-It is possible to assign a set of machining operations to a given finite set of machine tools.

We present our solutions in the report, as follows:

-a product description method,

-a conception of processing time evaluation,

-a some practical results of an implementation of the computer system.

INTRODUCTION

The designing of manufacturig processes, performed by process engineers with the use of traditional methods, is very labour-intensive and prone to errors resulting from the subjective human factor. The application of the electronic computation techniques for this purpose allows of:

- reducing the labour intensity of the design works,

- shortening the time required for the preparation of process documentation,
- improving the quality of manufacturing process design through the possibility of faster preparing of alternative versions,
- creating the conditions for decision objectivisation regarding the division of the manufacturig process into machining operations and steps, assigning operations to particular working stations, selection of machining tools and measuring instruments,
- attaining the uniform intensity of standards,
- relieving the designer of routine works, such as the calculation of machining allowances, definintion of machining parameters, etc.

The proposed system must include the following features:

- facility in data preparation,
- small quantity of input data,
- clear manner of their encoding,
- elasticity of the system,
- calculating precision within 10 - 15% of the standard manufacturing process.

It is pointed out that the increase in the assumed precision of calculation would require a considerable extension of the system, involving a rise in system operational costs. That is why consideration was given to the proportion of profits obtained through oncreasing the system precision and the system operational costs. This relation is shown in Fig. No. 1

As it follows from Fig. No. 1, starting with a certain point R / 2nd porlion of the curve / even a slight increase of calculatin precision /Δ P/ requires a considerable larger expense /ΔK_2/ then an identical increase of calculation precision /$\Delta P = \Delta S$/ in the first porlion of the curve.

The solution of the system was based upon a number of observations regarding the dependence of the working stations on the manufacturing process and design of machine components.

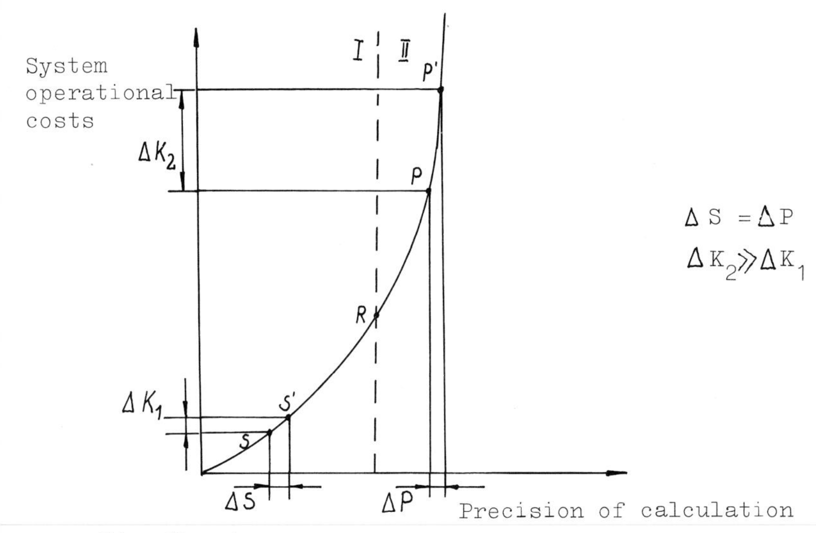

Fig. No. 1

These observations may be formulated as follows:

- the set of surfaces ancountered in machine components is finite and lends itself to classification,
- each machine component may be unequivocally described by the set of its surfaces,
- each surfaces may be machined in a finite number of operation sequences; these sequences for individual groups of surfaces lend themselves to classification,
- a given operation sequence provokes a specified division of machining allowances,
- there is a finite set of working stations suitable for the implementation of specified machining operations,
- there is a finite set of tools and fixtures suitable for the machining of the specified surfaces.

The idea of a system with thus specified assumptions consists in that the system takes over from a human the execution of all routine works and leaves only the definition of the machining concept itself.

The diagram of co-operation between a man and computer is shown in Fig. No. 2.

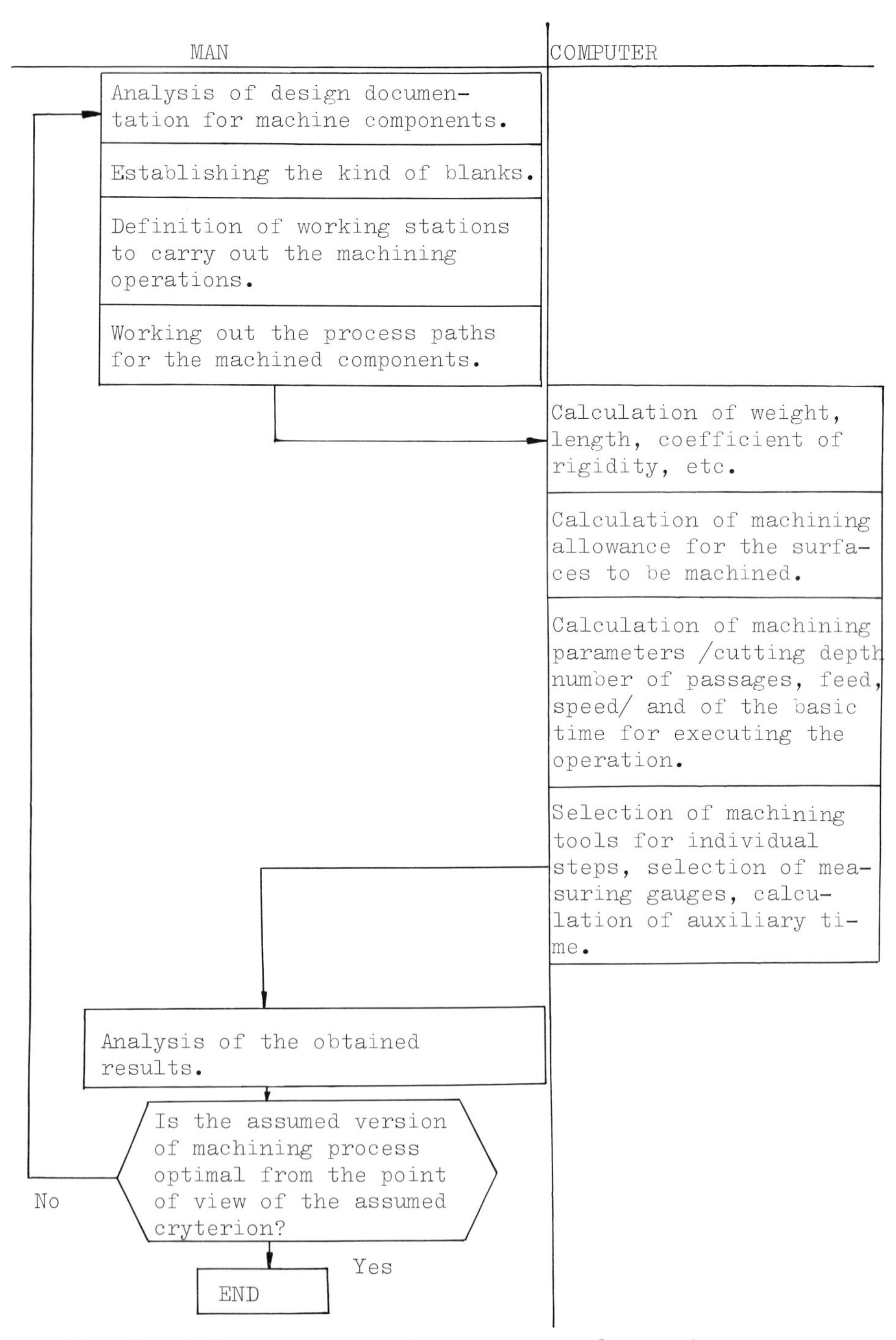

Fig. No. 2 Co-operation between a man and computer.

SYSTEM FUNCTIONING PRINCIPLES

The concept for the system construstion is shown in Fig. No. 3. The system is divided into a number of blocks, whose task is the implementation of its individual functions. Each of these blocks contains the indispensable permanent data base and a quasi-base constituted of mathematic formulas obtained from tabulated standards. From the diagram is omitted the reading block, and also the input data checking block, as well as the result emitting block. In the authors opinion, inclusion of these blocks in the diagram would detract from the clarity. The descriptions of individual block depicted on Fig. No. 3 is given in Table No. 1.

DESCRIPTION OF INPUT DATA

The input data may be divided into three groups:

1. Description of machine components.
2. Description of process routes.
3. Requirements regaring blanks and machining allowances.

ad1. The description of machine components has two parts. The first part is the general desription of the component containing the following information:

- description of the component shape,
- type of material /material group and sub-group, type of heat treatment, and type of blank/,
- yerly production program.

The second part contains the description of the elementary surfaces delimiting the bulk of component, in the following form:

- description of the shape of surface,
- accuracy of position as regards bases of reference,
- roughness class,
- tolerance class,
- surface dimensions /width, length height/,
- position of the surface in regard to reference bases and centre lines of the component.

ad2. The description of the process routing is composed from the

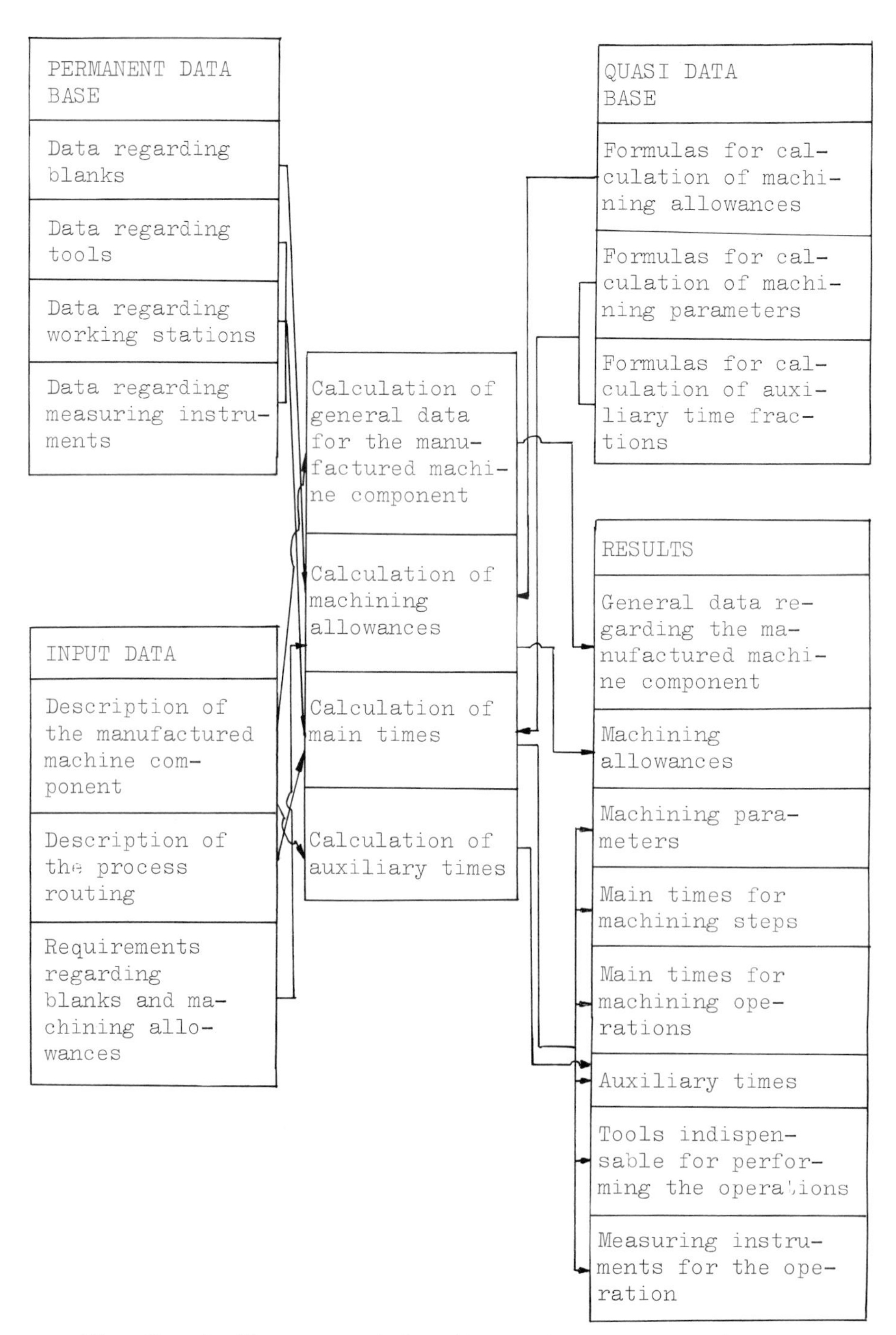

Fig. No. 3 The concept for the system construction

Table No. 1

BLOCK DESIGNATION	FUNCTION	PERMANENT DATA BASE	QUASI DATA BASE	INPUT DATA	RESULTS
Calculation of general data regarding machine components.	Calculation of such data as weight, rigidity and length of the component, etc.	-	-	Description of machine components.	General data of the components.
Calculation of machining allowance	Calculation of machining allowance for a surface with various types of blanks and various kinds of materials.	Data regarding blanks.	Formulas for machining allowance calculation.	Description of machine components. Requirements regarding machining allowance and blanks.	Machining allowance.
Calculation of times for the manufacture of machine components.	Distribution of allowance in accordance with the machining plan. Calculation of machining parameters for individual machining steps. Selection of machining tools. Calculation of main times for the machining steps and operations.	Data regarding	Formulas for the calculation of machining parameters.	Description of machine components. Description of the process routing.	Machining parameters. Main times for machining steps Main times for machining operations. Tools indispensable for performing the operations.
Calculation of auxiliary times.	Selection of measuring instruments. Calculation of auxiliary time fraction /time required for component mounting and removal, ect./	Data regarding working stations Data regarding measuring instruments.	Formulas for the calculation of auxiliary time fractions.	Description of machine components. Description of process routing.	Measuring instruments. Auxiliary times.

following items:

- vector of workstation types $ST=\{ST_j\}$ $1 \times N$
- matrix of operation steps $ZA=\{ZA_{kij}\}$ $R\times M\times N$
- matrix of regarding the machining tools $NA=\{NA_{kij}\}$ $R\times M\times N$
- matrix of data regarding the measuring instruments $PP=\{PP_{kij}\}$ $R\times M\times N$

where R = number of steps in an operation; M = number of elementary surfaces of the machine; N = number of operations.

$k = 1,\ldots,R$

$i = 1,\ldots,M$

$j = 1,\ldots,N$

ad3. The additional data regarding blanks contain the information regarding the directions of the process engineer regarding the diameters of normal blanks, or the increase of machining allowances for such blanks as a forging or casting.

DESCRIPTION OF RESULTS

The results of the system operation is an instruction sheet for a processing operation, including its parameters in the prescribed range for all desired combinations of input variables, and particulary:

- vector of component general data $DO=\{DO_l\}$ 1×11
- vector of machining allowances for the surface $N=\{N_i\}$ $1\times M$
- matrix of cutting speed $V=\{V_{kij}\}$ $R\times M\times N$
- matrix of the number of passages $I=\{I_{kij}\}$ $R\times M\times N$
- matrix of cutting depth $G=\{G_{kij}\}$ $R\times M\times N$
- matrix of feeds $P=\{P_{kij}\}$ $R\times M\times N$
- matrix of main times for steps $T=\{T_{kij}\}$ $R\times M\times N$
- matrix of main times for operations $TG=\{TG_{ij}\}$ $M\times N$
- matrix of auxiliary times $TP=\{TP_{ij}\}$ $M\times N$
- matrix of cutting tools $NN=\{NN_{kij}\}$ $R\times M\times N$

- matrix of measuring instruments $PR=\left\{PR_{kij}\right\}_{RxMxN}$

where R = number of steps in an operation; M = number of elementary surfaces of the machine; N = number of operations.

$k = 1,\ldots,R$

$i = 1,\ldots,M$

$j = 1,\ldots,N$

THE METHOD

The machining allowances, cutting parameters and fractions of the auxiliary time are established on the basis of the pertaining standards. The machining allowances depending upon the kind of blank are defined within the system as a function of the surface dimensions and overall dimensions of the componenet.

The cutting parameters in the majority of cases are the functions of surface dimensions and the required roughness and tolerance class, sometimes, however, they are functions of features which do not appear in the description of the machine component or its surfaces. This has a deciding influence upon the method used for the calculation of the main time. This is why there are two methods used by the system. One of these consists upon determination at the first stage of calculation of the cutting paramenters as a function of the dimensions, finish and tolerance class for the surface. Then these calculated cutting parameters are taken as a basis for the determination of the main time in accordance with the formula.The second method has been used in cases when the cutting parameters could be described exclusively as functions of surface parameters /beging also a function of other parameters not appearing in the formulas for the calculation of the main time/, or when they were the invoved function of these parameters. In such case, the results of simulation tests have been used for determination of the main time magnitude depending upon the magnitude of ots parameters.

It has been assumed that for every steps a sequence of auxiliary operations may be defined, depending upon the preceeding and successive steps.

The durations of such operations are a function of such conditions as:

- the size of the working space,
- the weight of the component,
- finishing clase, etc.

and have been described as mathematical functions.

In addition to the auxiliary operations related to the machining step, also as mathematical functions have been described the auxiliary times related to the machining operation. The unit times are calculated as the sum of the main time, the auxiliary time and the complementary time. The main time for the operation is considered as a sum of durations of individual machining steps, and the auxiliary time - as the sum of auxiliary times related to the machining step, increased by the magnitude of the auxiliary time related to the machining operation.

The complementary time is calculated as a percentage of the main time and the auxiliary time.

SUMMARY

Up to the present time has been worked out and programmed for the ODRA 1305 computer the module of the system as regards the manufacturing processes for the following component classes:

- shafts,
- discs,
- sleeves

machined on lathes and grinding machines in unitary, low-volume and medium-volume production.

In order to test the results obtained, statistical research has been carried out on a sample of 9 machine components requiring 60 machining steps for their manufacture. As it follows from this research, at assumed probability of 0.9 the expected value of error for the main time value shall be contained within the interval / +1.28%, +4.95% /, and for the auxiliary time value within the interval / +0.3%, +11.76% /. The results obtained confirm the hypothesis that the selected method for solution of the problem of computer-aided designing of manufacturing processes is correct.

At the present time the authors work on the succesive modules of the system pertaining to the remaining machining methods.

BIBLIOGRAPHY

1 W.D. Cwietkow, System automatyzacji projektowania procesów technologicznych / System for the automation of designing manufacturing processes /, PWN, Warszawa 1978.

2 M. Feld, Projektowanie procesów technologicznych typowych części maszyn / Designing of manufacturing processes for typical machine components /, N-T, Warszawa 1971.

3 S. Kunstetter, Narzędzia skrawające do metali / Metal-cutting tools /, N-T, Warszawa 1970.

4 J. Sikora, Optymalizacja procesów obróbki skrawaniem z zastosowaniem maszyn cyfrowych / Optimisation of metal-cutting processes with the use of digital computation aids /, N-T, Warszawa 1978.

5 J. Tymowski, Technologia budowy maszyn / Machine manufacturing processes /, PWN, Warszawa 1966.

6 A. Wąs, R. Izdebski, L. Kopiczyński, Automatyzacja projektowania procesów technologicznych w przemyśle maszynowym / Automation of designing the manufacturing processes in machine industry /, N-T, Warszawa 1971.

7 R. Wołk, Normowanie czasu pracy na obrabiarkach do obróbki skrawaniem / Standarisation of working times on metal-cutting machine-tools /, N-T, Warszawa 1972.

THREE INDUSTRIAL APPLICATIONS USING AN ELECTROLUX M.H.U. SENIOR ROBOT

E.A. Peirce & R.J. Grieve
Dept. of Production Technology
Brunel University, England

ABSTRACT

The first application is that of a vacuum handling system for transporting sheet material. Details of design and development of the vacuum system are discussed. The sequence of operations for moving sheet material from one work station to another is described along with the operational sequence for moving and stacking of cardboard boxes.

An interesting feature of this work is that of surface condition of work material. The surface texture and porosity of a work material determines the actual ultimate vacuum available within the gripper cup and the required vacuum flow from the actual system.

The second application is concerned with the design of variable diameter (8-80 mm) gripper for use in a machining cell. This work is concerned mainly with the major design principles and design criteria.

The effects of gripper clamping are considered for a range of work materials.

The final application is a robot-linked inspection station for automatic gauging of gudgeon pins for surface finish, diameter and taper. This automatic system makes a simple reject or accept decision on surface finish and taper but accommodates three tolerance bands on diameter.

A simple gripper is employed in handling the gudgeon pins between gauging and surface-finish stations. The system incorporates a ramp where components are initially collected but returned at a specific position along the ramp, representing final grading of the pins. The ramp serves also to provide a closed loop arrangement. In practice, collect and delivery stations would exist.

The operating sequence of this system is discussed.

INTRODUCTION

The M.H.U. senior is a point to point cylindrical configuration robot. It is of modular form, the degrees of freedom necessary for an application are obtained from a range of column/rotary/arm/wrist/linear and gripper units. The telescopic arm has a stroke of 1100 mm with 8 programmable stop positions. The rotating unit provides a 360° section of action and potential for rotating an optional number of revolutions and has 8 programmable stops. The 500 mm column has 6 programmable stops. All movements are pneumatically actuated and the stops are manually set.

Sequences are controlled by a microcomputer packaged into a programmable controller. This has standard 48 inputs and 32 outputs both expandable to 127. The controller has a 3K ram memory which is expandable to 5K ram. With a 3K ram memory capacity this gives around 200 programmable stops.

The success of robot applications depends to a great extent on the auxiliary equipment needed to help the robot perform the task. In particular the robot gripper is the point where the robot interacts with the environment. The design of a general purpose gripper has not yet been fully developed and instead the robots 'hand' has to be selected or designed specifically for a particular application or a range of applications.

The factors that need to be considered when selecting a gripper for a particular application have been discussed in the past and typical of these is the check list presented below (ref. 1)

FOR THE COMPONENT AND FOR ANALYSING HANDLING PATTERNS

FACTOR	REMARK
LOAD	dead weight, component weight
FORCES	push and pull forces, inertia and retardation forces, centre of gravity
DIMENSIONS	size, accessability
GEOMETRY	space, shape variation, shape changes, degrees of freedom
ORIENTATION	positioning, accuracy
TOLERANCES	deviation, reorientation
CONTACT SURFACES	gripping surfaces, friction versus material and environment, fragility
TIME	speed of movement, operation times.

FOR THE METHOD OF HOLDING AND ITS MEANS OF ACTIVATION

FACTOR	REMARK
GRIPPING METHOD	clamping, vacuum, magnet and combinations.
GRIPPING FORCE	force balance, sum of forces, control, surface pressure.
MATERIAL	rigidity, porosity, permeability, friction strength, fatigue strength
ENVIRONMENT	effect of oil, moisture, chemicals temperature
TIMES	gripping, releasing

OTHER FACTORS

FACTOR	REMARK
POSITION CHANGES	distortion, wear
FLEXIBILITY	exchangeability, standards, mattings, connections
PRODUCT DESIGN	modifications of the component
TIMES	design and fabrication
COSTS	design, construction, fabrication, assembly, test, running in
MAINTENANCE	service, spare parts

HE VACUUM HANDLING SYSTEM

Picking up, transporting and palletizing sheet materials is usually accomplished by one of two methods, namely magnetic or vacuum pick up systems. Magnetic grippers can work only on ferrous components. Vacuum grippers do not have this limitation, but they usually require a more complex system, and hence are more expensive. One advantage of gripping with magnets or vacuum is that only one surface is needed to pick up the component resulting in no inserts or other special provisions such as side clamps, chains, or slings being required. On some types of loads it can eliminate the need to use pallets.

The following checklist (ref. 2) of application factors given below, presents the criteria to be considered for each application, before purchase or design of a vacuum handling unit.

1. Size, weight, contour, condition, and porosity of the material to be handled.
2. Temperature of the material.
3. Accuracy of placement required.
4. Cycle time required to move the material from one place to another.
5. Ambient temperature range and other atmospheric conditions of the area in which the unit will operate.
6. Available floor space and headroom, heights of existing stacks, conveyors and skids.

The system developed for lifting sheet materials is shown in Fig. 1. A rotary vacuum pump capable of producing a maximum ultimate vacuum of 735 mm Hg with a suction capacity of 371 L/min was used in the system.

The air admittance valve is incorporated in the main line, and can be used to control the maximum ultimate vacuum within the system by allowing air in, thus reducing the vacuum. However, in a production system the air

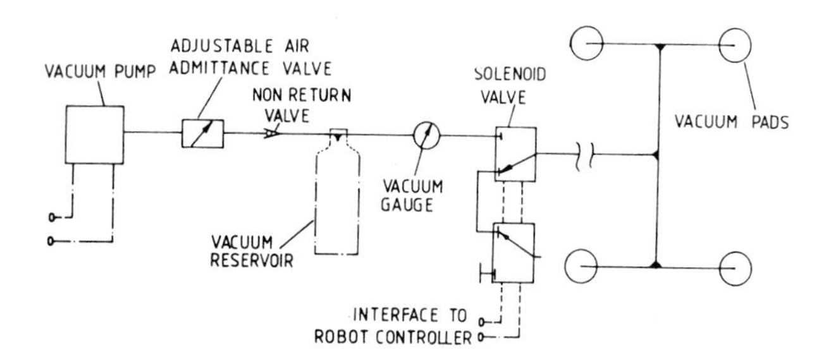

Fig. 1. Outline diagram of Vacuum Handling System

admittance valve would normally be in the closed position. In this particular development arrangement it was possible to determine the minimum ultimate vacuum necessary to lift a range of materials.

The non-return valve is incorporated in the main vacuum pump intake line to prevent losses through the pump in the event of a power or pump failure.

A vacuum reservoir is fitted as a "top up" system, with the pump supplying vacuum losses. In addition this reservoir acts as an emergency safeguard. i.e. in the event of power or pump failure the attitude of lift is maintained. The reservoir can also provide for a more efficient "pick-up" as the opening of the solenoid valves result in full flow across the valve. This can be far in excess of the pump size, so that a maximum impulse is obtained to effect the seal.

The solenoid valves connect the cups to the reservoir providing a switching facility to the cups when required. One of these valves switches the vacuum to the cups whilst the other is available in the event of a power failure thus maintaining a vacuum in the system.

Fig. 2 illustrates the lifting frame of the gripper together with the vacuum cups. Four cups (diameter 75 mm) were required in order to achieve the maximum lifting capability of the MHU robot (15 Kg).

At an extreme extension and carrying the maximum load (15 Kg) the deflection of the frame was found to be 3.8 mm.

In order to compensate for angular variation in the work surface, the vacuum cups were fitted into universal joint bearings. The angular movement provided by this bearing being approximately 22° included angle of movement of the cup, which provides compensation for any misalignment up to 11° angular variation in any plane.

Fig. 2. Vacuum Lifting Frame

Applications

The example simulates the movement of sheet material from one workstation to another. This operation is typical of a work cycle that would be performed in industry (press blanking, etc.). Fig. 3a shows a detailed block diagram of the work cycle and Fig. 3b the workstation and the robot arm path.

A second program (not reported here) simulated the transporting of cardboard boxes from one station and stacking at a second workstation. This type of work is common in flowline processing, typically with the food manufacturing industry.

Surface Texture and Porosity of Materials

The surface texture and porosity of the work material determines the actual ultimate vacuum available in the cup and the required vacuum flow from the vacuum system.

The ideal work materials are those that possess smooth air-tight surfaces.

Materials associated with a porous surface necessitate the use of larger area suction pads or an increase in vacuum in order to compensate for the system losses.

Some materials may look similar but in fact can exhibit very different boundary conditions between secure and unstable handling of the work.

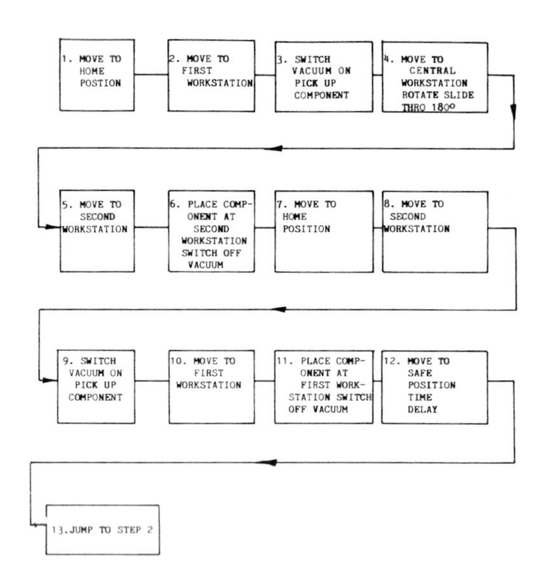

Fig. 3a. Sequence of operations for movement between Workstations

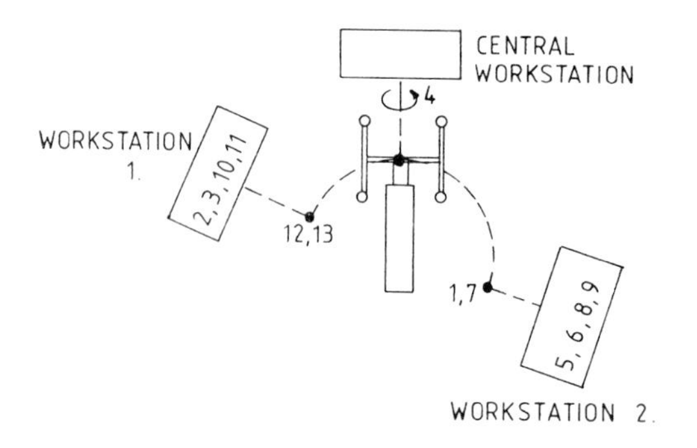

Fig. 3b. Workstation Layout and Sequence Steps

An electron microscope investigation reveals that materials such as metals provide the near ideal surface condition for maximum vacuum sealing. Materials such as plywood, can be handled with relative safety but blockboard, for example, exhibits a rough surface and has high porosity which makes handling difficult. Cardboard is associated with high porosity even though it exhibits a smooth surface. Thus the nature of the surface is an important factor when using a vacuum gripper system and robot accelerations and retardations need careful control if the surface results in the system being close to its lifting limit.

A GRIPPER FOR LIFTING VARIABLE DIAMETER COMPONENTS

The requirement was for a gripper to be designed in order to load cylindrical workpieces into an N.C. lathe; this combined unit forming part of a larger machining cell. The MHU senior robot is limited in some respects in that it does not have continuous path control but relies instead on 'stops' which must be set by hand. Thus the gripper has to hold the components in such a way that their centre line always coincides with the centre line of the chuck irrespective of the workpiece diameter. The range of diameters was from 8 mm to 80 mm.

Many of the grippers used in industry revolve around pivoted linkage mechanisms actuated by air cylinders. The maximum clamping force is designed to be applied in one fixed position, at the end of the cylinder travel. It would be difficult, using these methods, to achieve a constant clamping force over a range of positions.

Four bar linkages have been used to obtain parallel motion, and work has been conducted on a computer optimisation technique for determining the applicability of straight line producing four bar linkages. However, whilst the jaws close in a parallel mode, the height of the jaws drops according to the width of clamping. Thus their application is directed towards fixed dimensioned straight sided components.

To generate a straight line motion which is necessary in order to eliminate

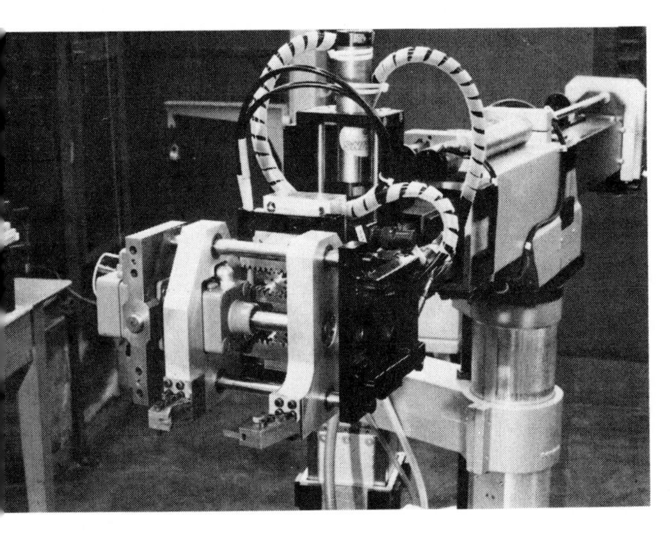

Fig. 4. The Variable Diameter Gripper

torsional effects on the grasped objects that would occur with curved or non-parallel finger closing trajectories, parallel rails were adopted. This offered a compact and accurate solution and made possible the application of the clamping force to an infinite number of positions along its travel.

A method then had to be devised to bring the two jaws together to meet accurately at the centre of the robot arm. In view of the fact that the MHU robot is pneumatically operated, it was decided to use pneumatic cylinders which would operate when called to do so from the robot control unit. An important consideration was that the gripper had to be operated in a manufacturing cell and thus spatial restrictions resulted when entering the machining area. Thus cylinders mounted externally to the jaws could cause problems.

One of the main problems of using cylinders is selecting the correct size and stroke, whilst preventing a bulky design. Thus a cylinder stroke length equal to the travel of one jaw was chosen and a method of transmitting the force to the opposite jaw had to be devised. A solution was found by using a rack and pinion. To ensure that a balanced force was transmitted to the opposite jaw, thus eliminating torsional effects that may prevent smooth, slow and positive motion, three racks and two pinions were used. The positioning of the cylinder was crucial in order to achieve adequate movement from such a precise stroke. (A 50 mm stroke length from an air cylinder was available, whilst 40 mm working length was required). A trunnion mounted cylinder acting through one jaw with clearance on the other was chosen, see Figs. 4, 5 and 6. To ensure that low speed movement was smooth and continuous, bearing mountings were vital. To this end, the pinions were mounted in roller ball bearings and the jaws operated on linear ball bushings.

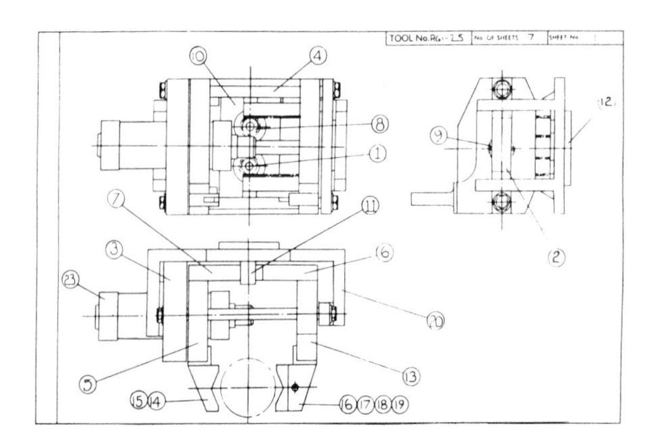

Fig. 5. Gripper Assembly Drawing

TOOL No. RG1-25	No. OF SHEETS 7	SHEET No 2

ITEM No	No OFF	DESCRIPTION	MATERIAL
1	2	SPUR GEAR SHAFT	KE 672
2	1	GUIDE MOUNTING	M S
3	1	TRUNNION MOUNTING	M S
4	2	GUIDE RAIL	KE 672 (HARDEN)
5	1	LHS GRIPPER SUPPORT	ALUMINIUM HE 15
6	2	RHS DRIVE RACK	MR20 20 500
7	1	LHS DRIVE RACK	MR20 20 500
8	2	SPUR GEAR	MA20 12
9	2	SPECIAL NUT	M S
10	1	DRIVE PLATE ASSY	M S
11	2	PRESSURE SUPPORT	KE 672 (HARDEN)
12	1	MOUNTING PLATE	M S
13	1	RHS GRIPPER SUPPORT	ALUMINIUM HE 15
14	1	LHS GRIPPER	KE 672 (HARDEN)
15	1	GRIPPER TOP PLATE	KE 672 (HARDEN)
16	1	RHS GRIPPER	KE 672 (HARDEN)
17	1	RHS CLAMP	KE 672 (HARDEN)
18	1	RHS GRIPPER TOP PLATE	KE 672 (HARDEN)
19	1	SPECIAL PIN	KE 672 (HARDEN)
20	1	MAIN FRAME	GCWS

ITEM No	No OFF	DESCRIPTION	MATERIAL
21	2	SPUR GEAR BEARING	RHP 6032
22	4	GUIDE RAIL BEARING	1525
23	1	AIR CYLINDER	M 13050
24	2	LIGHT SPRING	3.5 mm Ø × 15 mm Lg.

1) NOTE: THE HORIZONTAL ℄'S THRO' ITEMS 2,3,5,10 & 13 SHOULD HAVE A COMMON ℄ WHEN ASSEMBLED IN MAIN FRAME.

2) NOTE: THE ACCURATE ALIGNMENT THRO' TIGHT CONCENTRICITY OF THE ℄ OF THE 15 mm & 25 mm DIA HOLES IN ITEMS 3,5,13 & 2 IS VERY IMPORTENT, TO ENSURE THAT THE GUIDE RAILS (ITEMS 4) WILL BE PARALLEL TO EACH OTHER ON ASSEMBLEY.

GENERAL TOLERENCE ± 0,0635 MM UNLESS OTHERWISE STATED

Fig. 6. Material Schedule

The grasping force required from the robot depends on the weight of the part, the friction between the part and the fingers, the magnitude of accelerations and retardations and the relationship between the direction of movement and the position of the fingers on the part. In handling the part the worst situation is when the acceleration forces are parallel to the contact surfaces of the fingers. Here, friction alone has to hold the part. For movement in the horizontal plane inertia forces of 2g can be expected whilst in the vertical direction a magnitude of 3g may apply (ref. 3). These values have been quoted as 'typical' of robots and subsequent measurements on the MHU robot showed them to be reasonable. A component weight of 3 Kgs could be expected assuming a maximum steel component size of 80 mm diameter by 80 mm length. Thus considering the 'worst' case and assuming the coefficient of friction to be 0.15 the clamping force is 589 N.

In order to introduce a factor of safety a 50 mm diameter air cylinder was adopted capable of applying a force of 1025 N from the robot air supply.

An object of circular cross-section cannot be held properly by a two point grasp without an indeterminacy of position. The minimum number of clamping positions is therefore three. Thus one pin jaw and one vee shaped jaw would be suitable. However, there will be a slight variation of component centreline as the diameter changes. For this reason the gripping surfaces consisted of two 'vee' jaws, one of which was pivoted. To keep them as compact as possible interlocking mechanisms were designed.

In an effort to provide control over the air cylinder, a flow restriction valve, and a pressure adjustment valve together with a gauge were connected to the gripper before installing it.

The gripper has been designed to be a basic unit upon which other modules can be added. This is to satisfy the criteria of flexibility. Therefore, by simple redesign of the gripper jaws, the unit can pick-up components whose geometric shape, as well as size and weight, varies. The unit is not just suited to external gripping surfaces, but also to internal gripping applications.

THE GAUGING OF GUDGEON PINS

In this particular application the MHU robot is linked to an inspection situation where gudgeon pins are automatically gauged for surface finish, diameter and taper. The system has shown itself to be reliable from both the instrumentation and robotic placement aspects (ref. 4).

It must be pointed out that this particular application simply demonstrates the principles of such a system, in a production environment certain refinements would be necessary. For example, the robot employed is grossly over powered in terms of programmable capacity and size; also the traditional contacting method of moving a stylus over a surface is unacceptable for use in production.

Fig. 7 shows the general arrangement of the inspection layout.

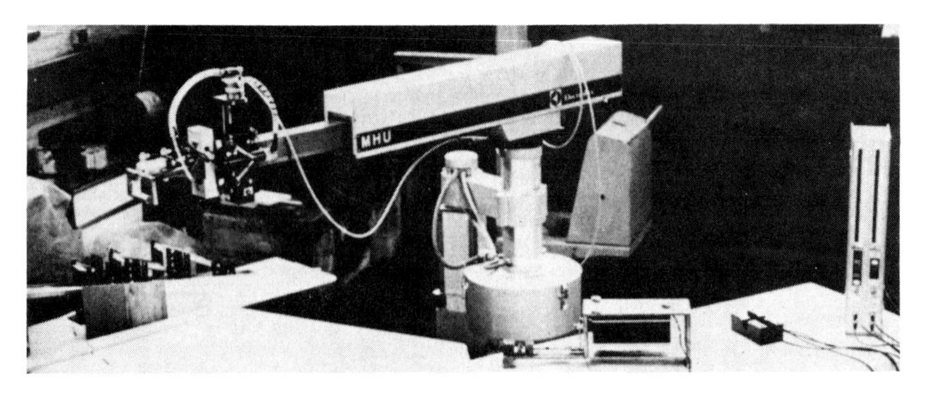

Fig. 7. General Arrangement of Inspection Layout

Inspection Requirements

In this application only outside diameter and taper are inspected. Acceptable diameter groups are:-

23.984 - 23.997 mm
23.997 - 24.000 mm
24.000 - 24.003 mm

Diameters outside these tolerance bands are rejected. Taper is measured between the two diameter readings and is to be within 2.5 microns over the inspection length. Surfaces less than 0.15 microns are accepted, those above 0.15 microns being rejected.

Fig. 8 illustrates the end effectors. The pins being picked up "end on", the gripper being provided with simple tapered "finger tips" to assist with component alignment into both inspection stations. This figure also shows the actual gauging station. This gauge employs inductive displacement transducers in conjunction with a solid state column display.

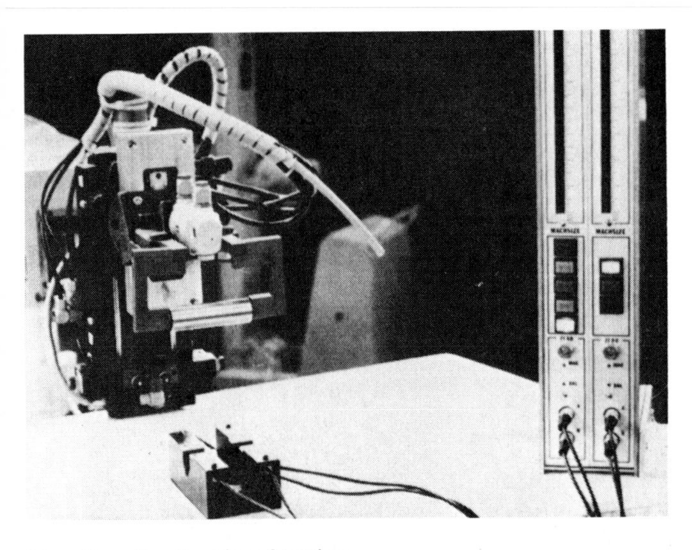

Fig. 8. The Gauging Station

The surface finish inspection station is shown in Fig. 9. The output signal for the inverted (to provide constant contacting pressure) roughness meter is rectified smoothed and then fed into the robot controller.

The gauging, surface finish and robotic device are linked by the robot controller. The mode of operation is simple in concept and is activated as follows. A master program is entered which responds to the electronic switches on the gauging and surface finish devices.

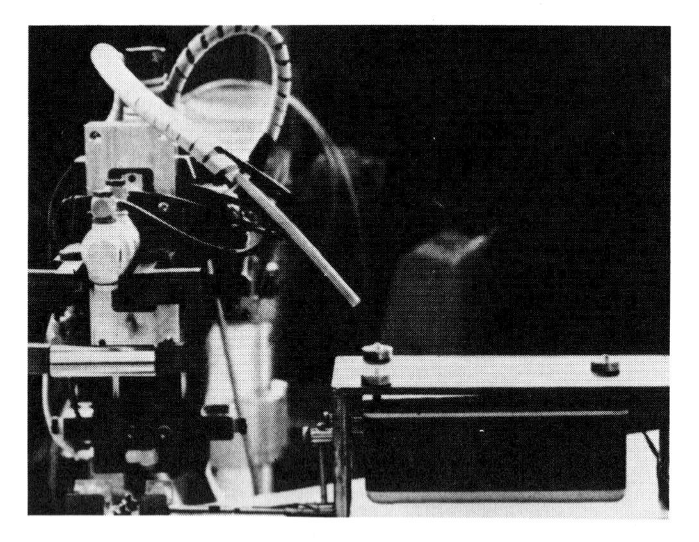

Fig. 9. Surface finish Station

Operating Sequence

Robot takes a gudgeon pin at the base of the ramp and places and releases it in the surface finish fixture, after inspection the pin is then removed and either placed in the gauging fixture or rejected. The diameter and taper are measured and depending upon the diametral tolerance band the pin is then positioned on the track, at the appropriate position. If the pin falls outside the tolerance band it is rejected. The robot then returns to the base of the track and picks up another component and the cycle is repeated.

The robot maintained a high level of placement accuracy and the instrumentation was found to be reliable with very little drift.

Although this work demonstrates the feasibility of such a system both the gauging and surface finish probes are of the contacting type. Although contacting probes are extensively used in industry, there are methods available (i.e. lasers) and under development that operate in a non-contacting mode.

REFERENCES

1. G. Lundstrom et. al., Industrial Robots - Gripper Review. International Fluidics Services Ltd., 1977.
2. E. Brooks, Use negative force for Positive Results, Design Engineering, April, 1973.
3. J.F. Engelberger, Robots in Practice, Kogan Page, London, 1980, p.43.
4. E.A. Peirce and R.J. Grieve, Robot Assisted Gauging of Gudgeon Pins, Sensor Review, July 1982, pp. 126-128.

PRINCIPLES AND APPLICATIONS OF SOLID STATE VISION

DR. W. NORTH[1], DR. T. PRYOR[2] and DR. W. PASTORIUS[3]
[1]Mechanical Engineering Dept., University of Windsor, Windsor, Ont., Canada
[2]Diffracto Ltd., Windsor, Ontario, Canada
[3]Diffracto Ltd., Windsor, Ontario, Canada

ABSTRACT

This paper describes basic electro-optical methods used to collect production quality information such as dimensions, missing operations, surface roughness, orientation, part location etc. Because it is an optical procedure, the part need not be contacted physically, the sensor can stand back from an otherwise messy production line and inspection at production rates of many parts per second is possible.

To demonstrate how these electro-optical solutions are translated into actual production con trol machinery, several examples are described and explained. These examples show what production quality information is being taken, how it is collected, how it is processed and/or used either as feedback or in real time quality control.

These examples are from turn key systems for in-line custom production gages and from standard products which either stand alone or are built into the production process by the customer.

INTRODUCTION

In today's marketplace the pressure of competition is even more important in the manufacturing industry as the economy is sinking or stagnant. Among others, competition requires that less scrap be produced, the product be more reliable, and that the products be less costly.

Cost is often governed by throughput and hence higher part rates are necessary to achieve the last objective. Reliability is better if 100% of throughput is inspected. Less scrap is produced if inspection takes place closer to or actually at the machines which produce the product.

The only hope to achieve all three goals economically is to inspect the parts automatically by some non contacting method. The most universal being an optical-electronic combination.

There are a host of optical solutions but no matter which one is chosen, it is generally not trivial to turn it into an inspection gage. Close to 100 companies now offer some form of visual inspection. Many offer a TV camera, either vidicon or solid state, coupled to a computer. While this often makes nice video display on an enhanced image, there is a significant expense and expertise required to make that into a working system which will replace an inspector for even a simple inspection task.

INSPECTION TASKS

The inspection task can be categorized into four modes as follows:

(a) dimension or position measurements

(b) missing or incomplete assembly

(c) flaw sensing

(d) surface roughness measurements

OPTICAL SOLUTIONS

Many principles are available ranging from interferometric methods which are the most accurate to simply detecting light intensity passing a station with a discrete detector, which is generally the least accurate. One method will not work for all situations, however some methods have general appeal.

Illumination is possible in a variety of concepts however incorrect use of illumination can easily defeat an otherwise effective solution. It is an art rather than a science and indeed a black art in many cases.

Dimensions or Position

While interferometric based methods are generally too sensitive for the factory environment, displacement or position can be monitored by several procedures.

(a) Project a spot of light onto a part and imaging this spot onto a dector array. Any movement of the part in the vertical sense, as indicated in figure 1, results in proportional movement of the imaged spot as measured by the detector.

(b) Imaging the profile of the part onto a detector array will also result in proportional change in the detector output as shown in figure 2.

(c) If clean lenses are difficult to ensure then diffracting coherent light through an aperture formed by a reference edge and the edge of the part itself is an alternative as shown in figure 3. This requires no lenses to implement however the part must be located more accurately.

Surface Flaws

In this instance light reflected from the illuminated surface of the part must be imaged onto a detector array as shown in figure 4. Incident light is scattered by the flaw resulting in a flaw image on the detector array.

Surface Roughness

If parallel light illuminates a machined surface, the surface roughness modulates the reflected light with characteristics which can be correlated to the surface roughness. See figure 5.

ELECTRONIC SOLUTIONS

Success or failure of an inspection procedure begins here as there are many transducers which can interpret the optical information.

Analog detectors are generally cheapest, sensitive and very fast however they suffer from stability, noise, and ultimately the accuracy with which their output transduces to position or dimension measurements.

Digital detectors in the form of column or two dimensional arrays are very effective devices which have stability, range and position resolution not available in analog detectors or a vidicon tube. The biggest disadvantage is their cost, dynamic range and the amount of information which must be taken and processed. However, single column arrays minimize these disadvantages and can be used in many applications.

HARDWARE/SOFTWARE

Combined hardware/software solutions such that only the necessary information need be stored and processed, are needed to meet part rates of several parts per second. Microprocessors can provide the necessary linearized answer from each sensor which are then managed by a host microprocessor to make part quality decisions, operate sorting gates, ring alarms, turn the inspection machine off after a run etc. etc.

The microprocessor also makes it easy to store tolerances for a wide range of parts, or to alter those tolerances or to omit a given inspection criterion and so on, without ever having to enter a printed circuit board to adjust a level controlling potentiometer.

Furthermore, the microprocessor can store statistical data and output it at any instant of time so that it can be used in real time to alter machine controls, or simply for production records. It may also be used to decide which department is making the scrap so that it can be charged back or returned and so on.

CONCLUSION

The slides demonstrate solutions to industrial quality control measurements using the technology described in the figures. The electro-optical components described, appear relatively straight forward and the parts can be purchased from several suppliers. There is however a significant amount of technology and black art associated with making these solutions work in a factory or even in an inspection room. Much of this technology can only be obtained by actually building a working system which includes integrating a mechanical handling package into the answer.

Experience demonstrates clearly the need for optimized lighting and optical systems. For example, how much optical magnification can be used effectively and alternatives to optical magnification. Illumination solutions are easy to propose but often difficult to incorporate into a good overall solution.

Finally, this experience is expensive, and the manufacturers of vision systems must be ready to absorb the cost up front in order to price their solutions to your problems competatively.

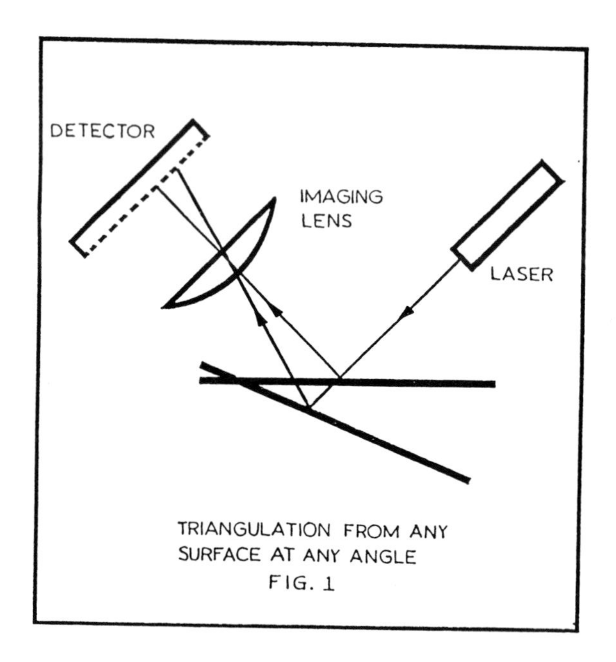

TRIANGULATION FROM ANY SURFACE AT ANY ANGLE
FIG. 1

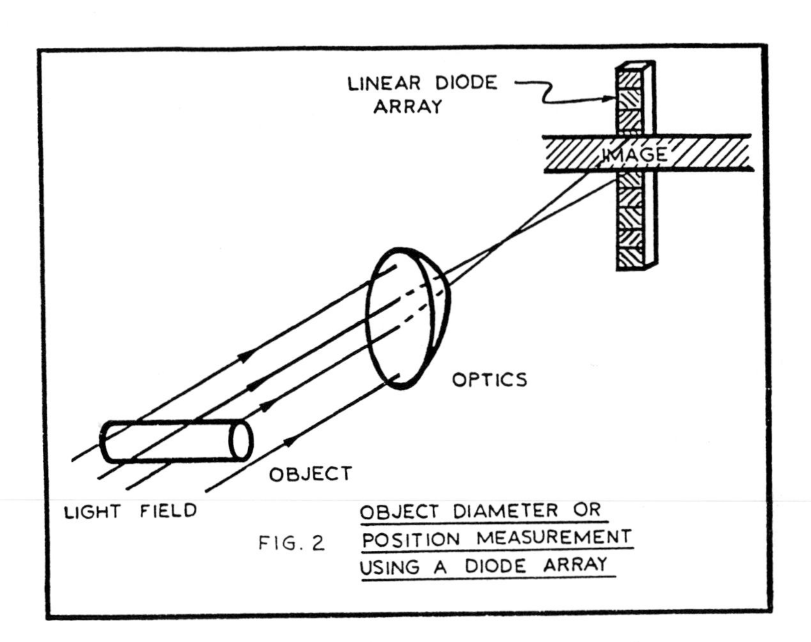

FIG. 2 OBJECT DIAMETER OR POSITION MEASUREMENT USING A DIODE ARRAY

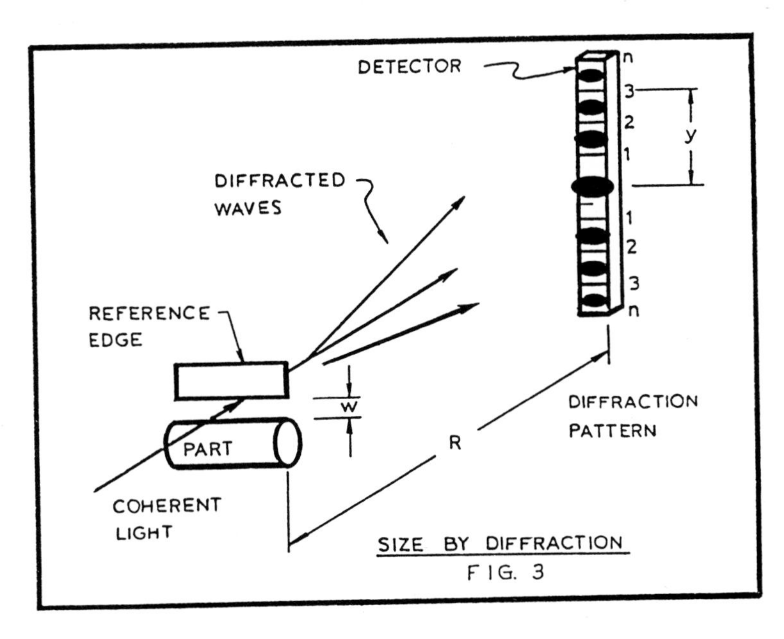

SIZE BY DIFFRACTION
FIG. 3

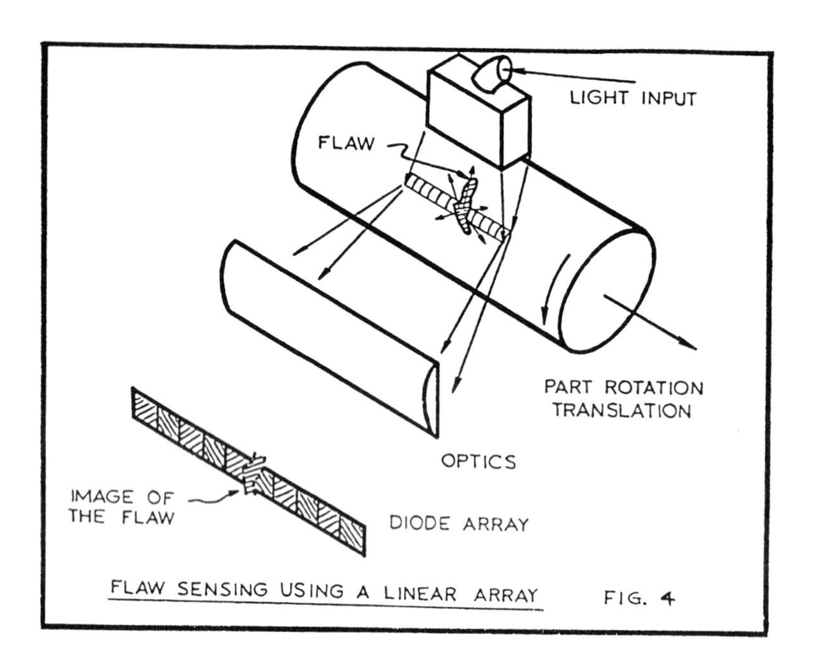

FLAW SENSING USING A LINEAR ARRAY FIG. 4

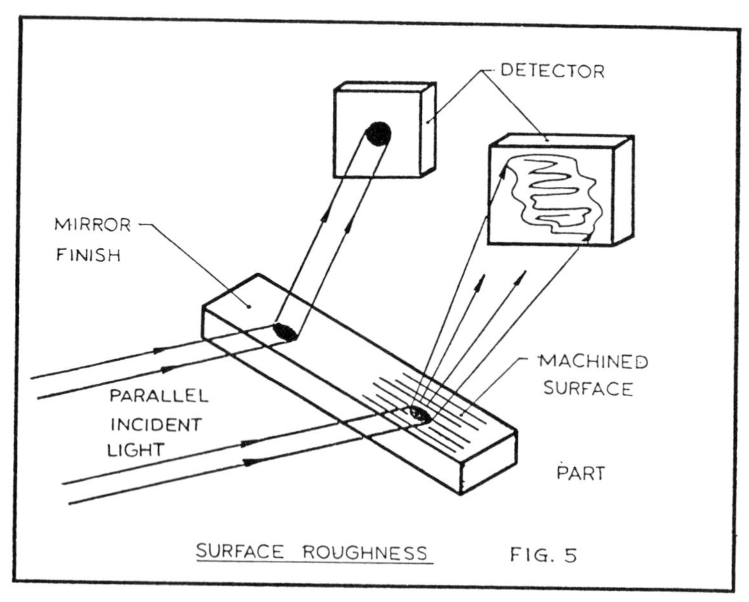

SURFACE ROUGHNESS FIG. 5

AN ALGORISM FOR SELECTING MANUFACTURING PROCESS VARIANTS FOR A SET OF COMPONENTS

Marek ZAMOJSKI, M.Sc. /Eng./
Warsaw Technical University
Institute of Organization and Management

ABSTRACT

The paper deals with the problem of selecting manufacturing process variants for a set of components according to a given criterion. This is a discrete problem which determines the level of industrial process automation.
For a given criterion a mathematical model and a calculating algorism is determined.
This is an integer programmnig optimization model, which is solved by a tree search algorism.
During the searching, special tests were made capable of reducing the number of iterations.
The described problem is exemplified by calculating procedure.

INTRODUCTION

Effective management involves the problems of appropriate choice between differents variants. One of the problems of this kind is the search of processes of machine components. Decisions made at this stage determine the volume of outlays required for obtaining a given production effect, and olso the prospective expenditure of exploitation of the production system.

Let it be assumed that a certain set of components is to be manufactured. For each component production process variants are designed, differing one from the another in the process of automation.

Variants should be found which fulfil the adopted criterion. By the criterion is uderstood the objective function and the constraint. The above problem is discrete in character.

ASSUMPTIONS

Concerning these variants, the following assumptions are made:

a/ All variants of the production processes of machine components lead to the same parameters of a particular machine component being obtained.

b/ The whole of the production program of a definite component is effected by one of the production process variants.

c/ The initial from of a particular machine component is identical for all of the production process of that particular component.

d/ Each of the production operations is characterized by the following numbers:
- number of the component involved in the particular operation,
- number of the production process variant,
- number of the operation, which at the same time is the number of the type of the workstand on which the operation is performed,
- duration of the operation.

e/ Production programs for all of the components of the set are determined.

THE CRITERION

For practical realization a certain definite combination of production process variants for all of the set's components are selected. Each of these combinations requires definite investment expenditure for production equipment. With each of these combinations is at the same time connected a definite number of employed workers.
It is worth investigating what minimal investment costs should be expended for the production of a definite set of components which a determined of workers. The real answer to this question is the finding an appropriate combination of production process variants of the set of components.

Thus the subjective function will be the investment costs and the constraint of the number of the employed.

In the set of all solutions is a solution requiring a certain smallest number of the employed and also a solution reguiring the largest number of those employed. The whole range - from the minimal to the maximal number of the employed should be - investigated this area. It is convenient to divide this section in to parts.

MATHEMATICAL MODEL

In this model to production operations there correspond vector elements, and to process variants, vectors. The set of process variants of a single machine component forms the production matrix of this components: there are as many of these matrixes as there are components.

The mathematical model is as follows:

Subjective Function

Minimalization of investment costs for the production equipment is expressed in the form

$$K = \min \sum_{k=1}^{k=s} y_k w_k + d_k, \tag{1}$$

where:

k - is the index of production equipment,

y_k - whole number of types k machines,

w_k - vector element of the purchase cost or of the selling price of type k machine,

w_k assumes values:

$y_k > 0$, then $w_k = K'_k$ if

$y_k < 0$, then $w_k = K''_k$

where:

K'_k - purchase cost of type k machine,

K''_k - selleing price of type k machine,

d_k - additional costs connected with a group of type k machines.

The variable y_k connected with the whole number of type k machines is in the form:

$$y_k = \sum_{i=1}^{i=m} \sum_{j=1}^{j=WM(i)} x_{ij} \eta_{ijk} - l_k \qquad k = 1,\dots,s \tag{2}$$

where:

i - index of the nzmber of the component,

j - index of the variant number of the production process,

WM(i) - number of production variants of component i,

m - number of components,

x_{ij} - variable connected with the choice of variant j of the production process of component i,

x_{ij} assumes values:

$$x_{ij} = 0 \quad \text{or} \quad x_{ij} = 1 \tag{3}$$

where:

$x_{ij} = 1$ - denotes the choice of production variant j of component i,

η_{ijk} - loading of the workstand k by the production process of variant j of component i,

l_k - number of type k workstands in workshop.

a/ limitation of the number of workers constraint

$$\sum_{k=1}^{k=s} \sum_{i=1}^{i=m} \sum_{j=1}^{j=WN\ i} X_{ij} Z_{ijk} \leqslant Z \tag{4}$$

Z_{ijk} - number of workers connected with realization of production process variant j of component i of operation k,

Z - total number of workers in the workshop, which cannot be exceeded.

b/ $$\sum_{j=1}^{j=WN(i)} X_{ij} \doteq 1 \qquad i = 1,\dots,m \tag{5}$$

Seeking Effective Production Processes

As is seen from the above, in this model there occur discrete, v, e, binary variables connected with the choice of a definite production process variant, and integral variables connected with number of machines in a particular type. The characteristic properties of this model determine it as the integral programmin problem.

ALGORISM

For the solution of this model, a calculating algorism has been designed, as well as its Fortran 1900 code representation on ODRA 1305 computer. This is a suboptimal algorism utilizing the controlled surveying technique. Special tests were used in this algorism. These tests results from the fact of the production matrix analysis having been performed prior to the actual performance of calculations. In this algorism the term of a machines type vector occurs. This is a binary vector connected with all machine types necessary for the whole set of production process variants of machine components. The binary structure of this vector in the course of calculations changes depending on testing results.

Algirism(Controlled Survey)

Step 1. Initial calculation. Calculations of the production matrix. Proceed to Step 2.

Step 2. Calculation of employment intervals. Find the maximal and minimal number of workers. Divide the workers employment range into intervals and determine their encompassing limits. Proceed to Step 3.

Step 3. Construction of branching test due to the interchangeability of machine types. Determine a set of technologically interchangeable machines types. Proceed to Step 4.

Step 4. Construction of branching test for a machine types minimum. Calculate the minimal number of machine types in the types vector. Proceed to Step 5.

Step 5. Construction of branching test for dependence of machine types. Find machine types which must occur in the case event of the absence of definite machine types. Proceed to Step 6.

Step 6. Branch generation according to machine types. Introduce zero on to the next successive position of the types vector of interchangeable machines. Proceed to Step 7.

Step 7. Test of the number of interchangeable machine types. If the current index of the machine types vector is greater than the number interchangeable machines, proceed to Step 12. Otherwise, proceed to Step 8.

Step 8. Test of the maximal number of zeros. If the number of zeros is equal to the maximal, proceed to Step 10. Otherwise, proceed to Step 9.

Step 9. Test of the dependence of machine types. Should the next successive interchangeable machine type not occur, proceed to Step 6. Otherwise, proceed to Step 10.

Step 10. Branch generation according to machine types. Introduce 1 into the next successive position of the wypes vector. Proceed to Step 11.

Step 11. Test of the number of interchangeable machine types. If the current interchangeable machine type vector numerator is greater than the number of interchangeable machine types, proceed to Step 12. Otherwise, proceed to Step 9.

Step 12. Elimination of the production process variants of machine components. Compare the generated machine types vector of each of the production process variants of all machine components. Check whether, using these machine types, you will be able to realize the current production process variant. If there exists even a single machine component unrealized by the use of vector, proceed to Step 13. Otherwise, proceed to Step 14.

Step 13. Backward motion. If no active apex occurs, proceed to Step 16. Otherwise, proceed to Step 10.

Step 14. Find solution to the linear programming task for a set of uneliminated production process variants of machine components. Proceed to Step 15.

Step 15. Optimality test. If the above solution is superior to the preceding one in a given employment range interval, memorize it and proceed to Step 13. Otherwise, without memorizing it, proceed to Step 13.

Step 16. Final. Write out a set of effective solutions.

EXAMPLE

The essential fraction of the algorism is Steps 6 to 11, in which generation tests of the binary type vector of the work stands are employed.

These are three of them, and their construction is best explained by example. Given is the production matrix of the production process variants of three machine components.

Machine component		1				2				3			
Interchangeability of machines	Variant / Machine	11	12	13	14	21	22	23	24	31	32	33	34
Z	M_1	η	η			η		η		η	η	η	
Z	M_2			η	η				η		η		η
Z	M_3	η		η				η				η	
NZ	M_4	η	η	η	η		η		η	η	η	η	η
NZ	M_5	η	η	η	η								
NZ	M_6		η	η	η	η	η	η	η	η	η	η	η
Z	M_7		η	η		η	η	η			η		
Z	M_8	η					η		η	η		η	
There does not occur a number of machines		3	3	2	4	5	4	4	4	4	3	3	5
Inclouding uninterchangeabls machines		1	0	0	0	2	1	2	1	1	1	1	1
Inclouding interchangeable machines		2	3	2	4	3	3	2	3	3	2	2	3
MAX		4				3				3			
MIN		3											

Successive tests are based on the observations utilized in the generation of the binary machine types vector, which are given below. Let it be noted that all of the machine types employed in the above production processes can be divided in two groups:

- machines indispensable - such without which none of the production process variants of even a single machine component can be realized; these shall be termed uninterchangeable machines - Symbol NZ,
- the other machines - these shall be termed interchangeable machines - Symbol Z.

Both machine groups are marked by these symbols in the "Interchangeability of Machines" column.

This piece of information permits unities contained in the elements corresponding to uninterchangeable machine types to be permanently introduced in to the binary machine types vector.

This is the first test for the generation of the branch search tree. The second test is based on the observation that there exists a certain maximal number of machine types without which at least one production process variant of each machine component is capable of being realized.

In the binary machine types vector, this number corresponds to the maximal number of zeros.

For machine component 1 in our example this can be realized without the use of four machine types - for component 2 without that of three; for component 3 also without that of three.

These numbers are written in the "Max" line. If we assumed that we attempted production of all machine components without four machine types, we would not be able to realize any single variant of the machine components 2 and 3. Hence it follows that we are capable of realizing production of all machine components without at most three interchangeable machine types. This result is written in the "Min" line. Thus the binary structure of the machine types vector can contain three zeros at most, which in the course of the branch tree generation is being currently checked. If this condition does occur, the other vector elements corresponding to interchangeable machines are bound to assume the value of 1. This permits a next successive reduction following the first test of the tree branching number. It is found that the number of these branches can still be reduced by the application of a third test. Its construction is based on the observation that if the binary machine types vector some of these types do not occur then, in order to realize at least one production process variant of each of the machine components some of the other must inevitably occur - that is some of the interchangeable machines set.

Figure 1 show a total set of the solutions of the previously presented methematical model. As is seen from Fig. 1, this solutions set has a discrete character. The set of solutions effective on account of the criterion above, obtained by algorism, is situated on the lower envelope of the solutions field. From the character of the above algorism it follows that effective solution can be achieved without analysing the entire solutions set.

Fig. 1. Diagrammatic interpretation of the solutions set.

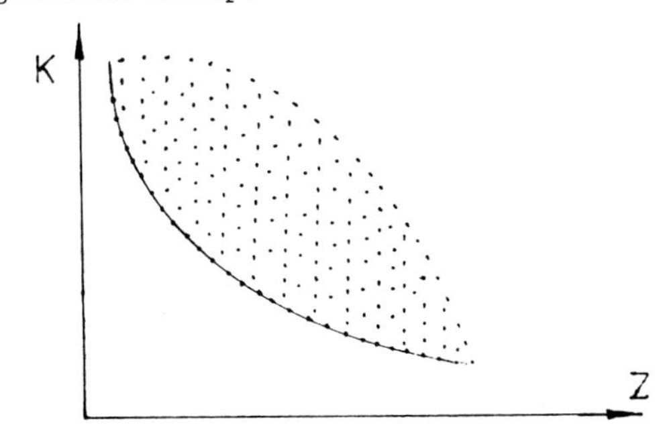

For the purpose of a clearer and fuller presentation of the algorism drawn, its most precise form possible from step 3 to step 11 is given.

Step 3. Branching test structure on account of interchangeable machines.
Step 3.1. Do to step 3.14 for KS = 1,...,S.
Step 3.2. Do to step 3.11 for ID = 1,...,M.

C KS is the control variable of machine type.
C S is the number of machine types.
C ID is the control variable of detail number,
C M is the number of details.

Step 3.3. Base LS3 = 0.

C LS3 is the numberator of operations performed on KS type machine.

Step 3.4. Do to step 3.8 for JW = 1,..., WN (ID) .

C Jw is the control variable of the variant number of the technological process of the detail.
C WN (ID) ist the vector of the number of technological process variants.

Step 3.5. See whether TAB(ID, JW, KS) = 0.0. If so, step 3.10, If not step 3.6.

C TAB is the matrix of technological process variants.

Step 3.6. Substitute LS3 = LS3 + 1 .
Step 3.7. See if LS3 = WN (ID). If so step 3.13. If not, contiunue step 3.4.
Step 3.9. Go to step 3.14
Step 3.10. Substitute PMS (KS) = 0 .

C PMS is the register of interchangeable and uninterchangeable machines.

Step 3.11. Continue step 3.2.
Step 3.12. Go to step 3.14 .
Step 3.13. Substitute PMS (KS) = 1 .
Step 3.14. Continue step 3.1.
Step 3.15. Substitute LMZ = 0 .

C LMZ to counter of interchangeable machines.

Step 3.16 Do to step 3.19 for KS = 1,...,S .
Step 3.17 See if PMS (KS) = 1. If so step 3.19, If not, step 3.18 .
Step 3.18 Substitute LMZ = LMZ + 1 .
Step 3.19 Continue step 3.16 .

Step 3.20 Substitute I = 0 .

C I is subsidiary control variable .

Step 3.21 Do to step 3.25 for KS = 1,...,S .

Step 3.22 See if PMS(KS) = 1 If so step 3.25. If not, step 3.23 .

Step 3.23 Substitute I = I + 1 .

Step 3.24 Substitute WKS2(I) = KS .

C WKS2(I) is register of the numbers of interchangeable machines.

Step 3.25 Continue step 3.22 .

Step 4. Branching test structure for the minimum of machines .

Step 4.1 Substitute L55 = 1000 .

C L55 is a numerator minimum of machines MIN .

Step 4.2 Do to step 4.20 for ID = 1,...,M .

Step 4.3 Do to step 4.4 for JW = 1,...,WN(ID) .

Step 4.4 Substitute WLZ(JW) = 0 .

C WLZ(JW) is the register of the number of interchangeable machines.

Step 4.5 Do to step 4.14 for JX = 1,...,WN(ID) .

Step 4.6 Do to step 4.11 for KX = 1,...,S .

Step 4.7 See if TAB(ID, JX, KX) = 0.0 IF so step 4.9. If not, step 4.8 .

Step 4.8 Go to step 4.17 .

Step 4.9 See if PMS(KX) = 1 If so, step 4.11 .

Step 4.10 Substitute WLZ(JX) = WLZ(JX) + 1 .

Step 4.11 Go to step 4.5 .

Step 4.12 Substitute L50 = 0 .

C L50 is the MAX numerator.

Step 4.13 do to step 4.17 for JX = 1,...WN(ID) .

Step 4.14 See if WLZ(JX) $\geqslant$ L50. If to, step 4.16. If not, step 4.15 .

Step 4.15 Go to step 4.17 .

Step 4.16 Substitute L50 = WLZ(JX) .

Step 4.17 Go to step 4.13 .

Step 4.18 See if L50 $>$ L55 If so, step 4.20. If not, step 4.19 .

Step 4.19 Substitute L55 = L50 .

Step 4.20 Go to step 4.2.

Step 5 Branching test structure on account of the dependence of machine types

C The result of operation of this part of the algorism is finding the matrix EL(I,J) .

C I is the successive number /not a real number/ of the interchangeable machine type.

C J is the index of the matrix columns .

C EL (I,J) assumesthe following values
0 or 1,...,LMZ or -1 .

C In the matrix EL it is written /registered/ what machine types of the set 1,...,LMZ must occur in the generated vector of WB types if the type I does not occur.

Step 5.1 Substitute L60 = 0 .
Step 5.2 Substitute L60 = L60 + 1 .
Step 5.3 Substitute J = 0 .
Step 5.4 Substutute J = J + 1 .
Step 5.5 See if J > LMZ.If so, step 5.8. If not, step 5.6.
Step 5.6 Substitute WPS1 (J) = 0 .
Step 5.7 Go to step 5.4.
Step 5.8 Substitute KS = WKS2 (L60) .
Step 5.9 Do to step 5.33 for ID = 1,...,M.
Step 5.10 Substitute I = 0 .
Step 5.11 Substitute I = I + 1 .
Step 5.12 See if I > LMZ If so, step 5.15.
Step 5.13 Substitute WPS I = 0 .
Step 5.14 Go to step 5.11.
Step 5.15 Do to step 5.26 for JW = 1,...,WN (ID)
Step 5.16 See if TAB (ID,JW,KS) > 0.0 If so, step 5.26.If not, step 5.17.
Step 5.17 Substitute L70 =0 .
Step 5.18 Substitute L70 = L70 + 1 .
Step 5.19 See if L70 > LMZ.If so, step 5.26. If not, step 5.20.
Step 5.20 See if TAB (ID, JW, WKS2 (L70)) = 0.0 If so, step 5.24.If not, step 5.21.
Step 5.21 See if WPS (L70) = -1 If so, step 5.16. If not, step 5.22.
Step 5.22 Substitute WPS (L70) = L70 .
Step 5.23 Go to step 5.28 .
Step 5.24 Substitute WPS L70 = -1 .
Step 5.25 Go to step 5.18 .
Step 5.26 Continue step 5.15 .
Step 5.27 Substitute J = 0 .
Step 5.28 Substitute J = J + 1 |
Step 5.29 See if J > LMZ.If so, step 5.33. If not, step 5.30

Step 5.30 See if WPS(J) < 0 If so, step 5.28. If not, step 5.31 .

Step 5.31 Substitute WPS1(J) = WPS(J) .

Step 5.32 Go to step 5.28 .

Step 5.33 Continue step 5.9 .

Step 5.34 Substitute J1 = 0 .

Step 5.35 Substitute J2 = 0 .

Step 5.36 Substitute L24 = 0 .

Step 5.37 Substitute J1 = J1 + 1 .

Step 5.38 See if J1 > LMZ.If so, step 5.44. If not, step 5.39.

Step 5.39 See if WPS1(J1) =0If so, step 5.37. If not, step 5.40.

Step 5.40 Substitute J2 = J2 + 1 ,

Step 5.41 Substitute EL(L60, J2) = WPS1(J1) .

C This means that if no type L60 machine occurs in the vector WB, a type WPS1 (J1) machine must then occur.

C Step 5.42. Substitute L24 = 1 ,

Step 5.43 Go to step 5.37 .

Step 5.44 See if L24 = 1 .If so, step 5.46. If not, step 5.45.

Step 5.45 Go to step 5.49 .

Step 5.46 Substitute J2 = J2 + 1 .

Step 5.47 Substitute EL(L60, J2) = -1 .

C -1 is the marker of the line end in the matrix EL .

Step 5.48 Go to step 5.50 .

Step 5.49 Substitute EL(L60, 1) = 0 .

C This means that if no type L60 machine occures, a machine of a different type need not occur.

Step 5.50 See if L60 = LMZ.If so, step 6. If not, step 5.51.

Step 5.51. Go to step 5.2.

C Steps 6.1 to 6.26 correspond to steps6 to 11 of the previously presented algorism outline, with the reverse order /sequence/ of the first and second test being adopted.

Step 6.1. Substitute LZ = 0 .

C LZ the zeros counter /minimum of machine types/ in the generated WB vector.

Step 6.2 Do to step 6.3 for J = 1,...,S.

Step 6.3 Substitute WB(J) = 0 .

C WB(J) element J of the generated vector of machine types.
It assumes the value of 1 if a given machine type occure or of 0 if it does not occur.

Step 6.4 Substitute I = 0 .

I is the numerator of the number of interchangeable machine types in the generated WB vector.

Step 6.5 See if LZ $\geqslant$ L55. If so, step 6.23. If not, step 6.6.

Step 6.6 Substitute I = I + 1 .

Step 6.7 See if I $>$ LMZ. If so, step 12. If not, step 6.8.

Step 6.8 Substitute J = 0 .

Step 6.9 Substitute J = J +1 .

Step 6.10 See if EL(I, J) $<$ 0 If so, step 6.5. If not, see if EL(I, J) = 0 . If so, step 6.18. If not, step 6.11.

Step 6.11 Substitute ELB = EL(I, J) .

Step 6.12 See if ELB $>$ I. If so, step 6.15. If not, step 6.13.

Step 6.13 See if WB(ELB) = 1 If so, step 6.15. If not, step 6.21.

Step 6.14 Go to step 6.21.

Step 6.15 Substitute L30 = J + 1 .

Step 6.16 See if EL(I, L30) = -1 If so, step 6.18. If not, step 6.17.

Step 6.17 Go to step 6.9 .

Step 6.18 Substitute WB (I) = 0 .

Step 6.19 Substitute LZ = LZ + 1 .

Step 6.20 Go to step 6.5.

Step 6.21 Substitute WB (I) = 1 .

Step 6.22 Go to step 6.5 .

Step 6.23 See if I = LMZ. If so, step 12. If not, step 6.24.

Step 6.24 Substitute I = I + 1 .

Step 6.25 Substitute WB (I) = 1 .

Step 6.26 Go to step 6.23.

REFERENCES:

1 Fiszel H., Theory of Investment Effectiveness and Its Application, PWN, Warszawa 1969.

2 Garfinkel R.S., Memhauser G.L., Integer Programming, PWN, Warszawa 1979.

3 Piekutowski J., Modernization of production Equipment of Machine Industry Plants, Mechanik, No 9/1977.

4 Piekutowski J., Zamojski M., Research work entitled: Analysis of Concepts and Algorisms of Selection of Effective Production Process Variants: Conducted at the Institute of Organization and Managament, PW, 1981.

AUTHOR INDEX

SUBJECT INDEX